MEMOIRS
of the
American Mathematical Society

Number 457

Multiplicative Homology Operations and Transfer

Norihiko Minami

November 1991 • Volume 94 • Number 457 (third of 4 numbers) • ISSN 0065-9266

American Mathematical Society
Providence, Rhode Island

1980 *Mathematics Subject Classification* (1985 *Revision*).
Primary 55P47, 55P91, 55S12.

Library of Congress Cataloging-in-Publication Data

Minami, Norihiko, 1958–
Multiplicative homology operations and transfer/Norihiko Minami.
p. cm. – (Memoirs of the American Mathematical Society, ISSN 0065-9266; no. 457)
"November 1991, volume 94 . . . third of 4 numbers."
Includes bibliographical references.
ISBN 0-8218-2518-6
1. Infinite loop spaces. 2. Homology theory. 3. Burnside rings. I. Title. II. Series.
QA3.A57 no. 457
[QA612.76]
510 s–dc20 91-28757
[514′.23] CIP

Subscriptions and orders for publications of the American Mathematical Society should be addressed to American Mathematical Society, Box 1571, Annex Station, Providence, RI 02901-1571. *All orders must be accompanied by payment.* Other correspondence should be addressed to Box 6248, Providence, RI 02940-6248.

SUBSCRIPTION INFORMATION. The 1991 subscription begins with Number 438 and consists of six mailings, each containing one or more numbers. Subscription prices for 1991 are $270 list, $216 institutional member. A late charge of 10% of the subscription price will be imposed on orders received from nonmembers after January 1 of the subscription year. Subscribers outside the United States and India must pay a postage surcharge of $25; subscribers in India must pay a postage surcharge of $43. Expedited delivery to destinations in North America $30; elsewhere $82. Each number may be ordered separately; *please specify number* when ordering an individual number. For prices and titles of recently released numbers, see the New Publications sections of the NOTICES of the American Mathematical Society.

BACK NUMBER INFORMATION. For back issues see the AMS Catalogue of Publications.

MEMOIRS of the American Mathematical Society (ISSN 0065-9266) is published bimonthly (each volume consisting usually of more than one number) by the American Mathematical Society at 201 Charles Street, Providence, Rhode Island 02904-2213. Second Class postage paid at Providence, Rhode Island 02940-6248. Postmaster: Send address changes to Memoirs of the American Mathematical Society, American Mathematical Society, Box 6248, Providence, RI 02940-6248.

10 9 8 7 6 5 4 3 2 1 95 94 93 92 91

Table of contents

ABSTRACT

Here we give a completely new treatment of the homology operations on QS^0. Our new idea concentrates on the Burnside ring of elementary abelian p-groups: To study an element $x \in H_*(QS^0)$, we first find some $y \in H_*(B(\mathbb{Z}/p)^r)$ and $a \in A((\mathbb{Z}/p)^r)$ such that $\alpha_{(\mathbb{Z}/p)^r}(a)_*(y) = x$, then we investigate a rather than x itself.

This approach should be the most advantageous and especially powerful when we study multiplicative homology operations. For example, we proved the odd primary multiplicative homology operations formula claimed by Tsuchiya (and whose gap questioned by May). In fact our estimates are sharper than Tsuchiya's original claim and also Madsen's mod-2 analogy. This leads to the following multiplicative transfer theorem for SG:

Theorem 3.21. (*Priddy when* p=2) *Let* δ *be the composite*:

$$B\mathbb{Z}_p \int \mathbb{Z}_p \to B\Sigma_{p2} \xrightarrow{D\text{-}L} Q_{p2}S^0 \xrightarrow{*[1-p^2]} Q_1S^0 = SG.$$

Then there is a map t: $SG \to Q(B\mathbb{Z}_p \int \mathbb{Z}_p)$ *such that*

$$SG \xrightarrow{t} Q(B\mathbb{Z}_p \int \mathbb{Z}_p) \xrightarrow{Q(\delta)} QSG \xrightarrow{\xi} SG$$

is an equivalence at p, *where* ξ *is the map induced by the infinite loop space structure of* SG.

1980 Mathematics Subject Classification: 55 P47, 55 P91, 55 S12.

Keywords and phrases: Infinite loop space, homology operations, SG, Burnside ring.

INTRODUCTION

The purpose of this paper is to present a totally new approach to homology operations on QS^0 — including their foundations and some applications. Some readers might say "Homology operations on QS^0 can be computed using the well known formulas such as (Mixed) Adem formula, (Mixed) Cartan formula, May's formula or Nishida relations. What is left?" Well, he's right in the sense that they can be computed in principle. But as far as the main tool of algebraic topology is calculation, "Can be computed in principle" alone is never satisfactory. The most typical example in our case would be the problem between Tsuchiya and May (cf. p.141 of [5]). The problem there comes from the complicated nature of the binomial coefficients, which can be thoroughly forgotten in our approach.

Our approach is as follows: For any $x \in H_*(QS^0)$, which we want to study, we find some $B(\mathbb{Z}/p)^r$, classifying space of elementary abelian group $(\mathbb{Z}/p)^r$, some $y \in H_*(QS^0)$, and some $b \in A((\mathbb{Z}/p)^r)$, the Burnside ring of $(\mathbb{Z}/p)^r$, such that

$$\alpha_{(\mathbb{Z}/p)^r}(b)_*(y) = x \in H_*(QS^0) \quad \text{(see Section 1 for } \alpha_{(\mathbb{Z}/p)^r}).$$

Even though we want to know $x \in H_*(QS^0)$, our study will be concentrated on $b \in A((\mathbb{Z}/p)^r)$ rather than x or y. The Barratt-Priddy-Quillen theorem [20] suggests that our approach holds on to the categorical (=Burnside ring) setting until the very last moment when we return to the homology level, whereas the classical approach held on to the categorical (=Burnside ring) setting only when they obtained the various formulas such as (Mixed) Adem formula, (Mixed) Cartan formula, and May's formula, and went into the complicated homology level immediately.

Aside from the affirmation of Tsuchiya's claim [26] (thus ending the problem

[1]Received by the editor July 21, 1988.

Received in revised form January 7, 1991.

between Tsuchiya and May), one of our main new results is the following:

Theorem 3.21. *(Priddy [21] for p=2) Let δ be the composite:*

$$B\mathbb{Z}_p\int\mathbb{Z}_p \to B\Sigma_{p2} \xrightarrow{D\text{-}L} Q_{p2}S^0 \xrightarrow{*[1-p^2]} Q_1S^0 = SG.$$

Then there is a map t: $SG \to Q(B\mathbb{Z}_p\int\mathbb{Z}_p)$ such that

$$SG \xrightarrow{t} Q(B\mathbb{Z}_p\int\mathbb{Z}_p) \xrightarrow{Q(\delta)} QSG \xrightarrow{\xi} SG$$

is an equivalence at p, where ξ is the map induced by the infinite loop space structure of SG.

Of course this is the multiplicative analogue of the prestigious Kahn-Priddy theorem [10] and the mod-2 case was already proved by Priddy [21] in 1977. But to give a homological proof of this for p=odd, Tsuchiya's claim is not sharp enough. Very fortunately our method allows us to prove a more enriched form of Tsuchiya's claim and a corresponding mod-2 version Madsen's result in Corollary 3.15:

Corollary 3.15. *Let $I=(J,K)$, $l(K)\geq 2$. Then*

$$\tilde{Q}_J(x_K) \equiv x_I + \sum b_L x_L \quad \text{mod.}\begin{cases} \bar{B}\#\bar{B} & \text{when } p=2 \\ \widetilde{\mathfrak{P}}(I_1)\bigcup\bar{B}\#\widetilde{\mathfrak{P}}(I_1)\bigcup\bar{B}\#\bar{B} & \text{when } p=\text{odd}, \end{cases}$$

where L runs over the admissible sequence with $l(K)\leq l(L)<l(I)$ and furthermore over $(0,a)$ $a\in\mathbb{N}$ when $p=2$ and $l(K)=2$.

In both Madsen's result (when $p=2$) and Tsuchiya's claim (when p=odd) the congruence had to be taken mod. $I_1\#I_1$. Like this, our approach gives many sharp results which interrelate the additive structure and the multiplicative structure, using p-adic exponential and p-adic logarithm at the Burnside ring level. Maybe what is most interesting is the fact that once a result of Tsuchiya [26] (Proposition 3.12) that the odd primary additive and multiplicative p-th powers are the essentially same is admitted, our study shows the odd primary cases are easier than the mod-2 case in the study of multiplicative homology operations because of the nature of the p-adic exponential homomorphism. The author does believe that the essential difficulty of the odd primary multiplicative operation exists in this fact, whereas Madsen's mod-2 analogue is rather simple (see Remark 3.13). The author hopes these would shed some light on the complicated multiplicative structure, which is very important in connection with the surgery through spaces like G/O. Our application to multiplicative homology operations is given through a homological proof of Theorem 3.21, since any homological proof of a transfer theorem requires a thorough knowledge of homology operation action.

Now the organization of this paper is as follows: In Section 1, we recall the map

$$\alpha_G\colon A(G) \to [BG, QS^0]$$

is a ring homomorphism and commutes with the (additive and multiplicative) transfers. (see Section 1 for the precise definition) This enables us to use the Burnside ring as a main tool to study homology operations. In Section 2, we give many calculational results of the Burnside ring of elementary abelian groups. Here are given the definitions of p-adic exponential exp_p and p-adic logarithm log_p, and the various categorical (=Burnside ring) correspondence to some important homology elements we want to know. In Section 3, we give some sharp calculations of homology operations such as Corollary 3.15 and prove the above cited Theorem 3.21, the multiplicative analogue of the Kahn-Priddy theorem.

Even though our results presented here are just for QS^0, it is apparent that our method suggests the following: Whenever we study homology operations on an infinite loop space, we have to find its category model and (homology) detecting subcategories and play around at that subcategory level until the very last moment when we return to the homology level.

This material is somewhat a proof-expanded version of the author's thesis [27] at Northwestern Universiry written under the direction of Professor Stewart Priddy. The autor expresses his gratitude to the refree for demanding him to supply more details of the proofs. Therefore the reader should also thank the refree. The author also expresses his gratitude to Professorsers Dan Kahn, Mark Mahowald, Peter May, Goro Nishida, Stewart Priddy, and Masahiro Sugawara. Maybe the author's wife Noriko desearves his gratitude too, especially for preparing movings packings 4 years in a row since [27] was first typed. Finally he thanks finantial supports by Educational Project for Japanese Mathematical Scientists for its travel support, NSF grant no. DMS-8641403 at Northwestern, and NSF grant no. DMS-8505550 at MSRI where this paper was revised.

§1. MULTIPLICATIVE TRANSFER OF THE BURNSIDE RING AND QS^0

Let $A(G)$ be the Burnside ring of a compact Lie group G defined by tom Dieck [6]; $A(G)$ is the set of equivalence classes of compact G-manifolds with the equivalence relation

$$X \sim Y \quad \Leftrightarrow \quad \chi(X^H) = \chi(Y^H) \text{ for any } H \subseteq G.$$

(In fact the most general definition includes compact G-ENRs [6, 5.5.] or finite G-spaces [11], but our interest here is just manifolds.). Addition is given by disjoint union and multiplication is given by cartesian product. Furthermore there are two kinds of transfers in the Burnside ring [6, 5.12. and 5.13];

$$\mathrm{Ind}^{\oplus G}_{H}\colon A(H) \longrightarrow A(G), \quad X \mapsto G \times_H X$$

$$\mathrm{Ind}^{\otimes G}_{H}\colon A(H) \longrightarrow A(G), \quad X \mapsto \mathrm{Hom}_H(G, X),$$

where H is a closed subgroup of G, and in the latter we require H to be finite index in G. As these notations suggest, $\mathrm{Ind}^{\oplus G}_{H}$ preserves the addition and $\mathrm{Ind}^{\otimes G}_{H}$ preserves the multiplication. When G is finite, tom Dieck's definition is the same as the original definition i.e. the Grothendieck ring of finite G-sets which is the free abelian group with basis $\{G/H\}$. But even in this case, if we stick to the original definition, we find much difficulty in defining and evaluating $\mathrm{Ind}^{\otimes G}_{H}$ on elements whose expression contains formal inverses; tom Dieck's definition, which is a set of equivalence classes not merely a formal free abelian group, alleviates this problem.

On the other hand our favorite space QS^0 is an E_∞ ring space [14] [11]. So there are two kinds of products

$$\theta\colon QS^0 \times QS^0 \longrightarrow QS^0$$

$$\xi\colon QS^0 \times QS^0 \longrightarrow QS^0,$$

which correspond to the loop sum (=addition) and the composition

(=multiplication) respectively, and these maps admit an extension to the operad action

$$\theta\colon E\Sigma_n \underset{\Sigma_n}{\times} (QS^0)^n \longrightarrow QS^0$$
$$\xi\colon E\Sigma_n \underset{\Sigma_n}{\times} (QS^0)^n \longrightarrow QS^0,$$

so that θ corresponds to an additive operad (e.g. Steiner's operad [23][14, 6.7.]) action and ξ corresponds to a multiplicative operad (e.g. linear isometry operad [14]) action. In fact QS^0 is an infinite loop space with the loop sum and SG $(=Q_1S^0)$ is an infinite loop space with the composition product, and up to homotopy (since this depends on the way how we take $E\Sigma_n$) θ and $\xi|_{E\Sigma_n \underset{\Sigma_n}{\times} (SG)^n}$ may be thought as the Dyer Lashof maps of these infinite loop spaces respectively (see the comment after the proof of Key Fact below). The transfers of the stable cohomotopy theory for finite coverings are defined using these.

Now the purpose of this section is to recall the map

$$\alpha_G\colon A(G) \longrightarrow [BG,\, QS^0] = \pi_s^0(BG_+),$$

for each compact Lie group G, and note that this preserves the structures of $A(G)$ and QS^0 mentioned above:

Lemma 1.1. α_G *is a ring homomorphism: For any compact G-manifolds X and Y,*

$$\alpha_G(X+Y) \sim \theta\circ(\alpha_G(X)\times\alpha_G(Y))\circ B\Delta$$
$$\alpha_G(X\times Y) \sim \xi\circ(\alpha_G(X)\times\alpha_G(Y))\circ B\Delta$$

i.e. up to homotopy, we have the following factorizations;

$$\alpha_G(X+Y)\colon BG \xrightarrow{B\Delta} BG\times BG \xrightarrow{\alpha_G(X)\times\alpha_G(Y)} QS^0\times QS^0 \xrightarrow{\theta} QS^0$$

$$\alpha_G(X\times Y)\colon BG \xrightarrow{B\Delta} BG\times BG \xrightarrow{\alpha_G(X)\times\alpha_G(Y)} QS^0\times QS^0 \xrightarrow{\xi} QS^0.$$

Lemma 1.2. α_G *is (both additive and multiplicative) transfers commuting: Let H be a closed subgroup of G such that $|G/H|=n$ and let $\Psi\colon G \longrightarrow \Sigma_n\int H$ be the usual* *homomorphism* [8][10], *which is defined uniquely up to conjugacy. Then for any compact H-manifold X,*

$$\alpha_G(\mathrm{Ind}^{\oplus}{}^G_H(X)) \sim \theta \circ \Big(1 \underset{\Sigma_n}{\times} (\alpha_H(X))^n\Big) \circ B\Psi$$

$$\alpha_G(\mathrm{Ind}^{\otimes}{}^G_H(X)) \sim \xi \circ \Big(1 \underset{\Sigma_n}{\times} (\alpha_H(X))^n\Big) \circ B\Psi$$

i.e. up to homotopy, we have the following factorizations:

$$\alpha_G(\mathrm{Ind}^{\oplus}{}^G_H(X)) \colon BG \xrightarrow{B\Psi} B(\Sigma_n \textstyle\int H) \xrightarrow{\iota} E\Sigma_n \underset{\Sigma_n}{\times} (BH)^n \xrightarrow{1 \underset{\Sigma_n}{\times} (\alpha_H(X))^n} E\Sigma_n \underset{\Sigma_n}{\times} (QS^0)^n \xrightarrow{\theta} QS^0$$

$$\alpha_G(\mathrm{Ind}^{\otimes}{}^G_H(X)) \colon BG \xrightarrow{B\Psi} B(\Sigma_n \textstyle\int H) \xrightarrow{\iota} E\Sigma_n \underset{\Sigma_n}{\times} (BH)^n \xrightarrow{1 \underset{\Sigma_n}{\times} (\alpha_H(X))^n} E\Sigma_n \underset{\Sigma_n}{\times} (QS^0)^n \xrightarrow{\xi} QS^0.$$

Lemma 1.3. *For any natural number n and a compact Lie group H, define the additive and multiplicative power operations as follows;*

$$\mathfrak{P}^n_{\oplus} \colon A(H) \longrightarrow A(\Sigma_n \textstyle\int H), \quad X \mapsto nX$$

$$\mathfrak{P}^n_{\otimes} \colon A(H) \longrightarrow A(\Sigma_n \textstyle\int H), \quad X \mapsto X^n,$$

where this X^n has $\Sigma_n \int H$ action [6, 5.13.7.]

$$(\sigma; h_1,\ldots,h_n)(x_1,\ldots,x_n) = (h_1 x_{\sigma^{-1}(1)},\ldots,h_n x_{\sigma^{-1}(n)})$$

and $\Sigma_n \int H$ acts on nX similarly.

Then we have

$$\alpha_{\Sigma_n \int H}(\mathfrak{P}^n_{\oplus}(X)) \sim \theta \circ \Big(1 \underset{\Sigma_n}{\times} (\alpha_H(X))^n\Big)$$

$$\alpha_{\Sigma_n \int H}(\mathfrak{P}^n_{\otimes}(X)) \sim \xi \circ \Big(1 \underset{\Sigma_n}{\times} (\alpha_H(X))^n\Big)$$

i.e. up to homotopy, we have the following expressions;

$$\alpha_{\Sigma_n \int H}(\mathfrak{P}^n_{\oplus}(X)) \colon B(\Sigma_n \textstyle\int H) \xrightarrow{\iota} E\Sigma_n \underset{\Sigma_n}{\times} (BH)^n \xrightarrow{1 \underset{\Sigma_n}{\times} (\alpha_H(X))^n} E\Sigma_n \underset{\Sigma_n}{\times} (QS^0)^n \xrightarrow{\theta} QS^0$$

$$\alpha_{\Sigma_n \int H}(\mathfrak{P}^n_{\otimes}(X)) \colon B(\Sigma_n \textstyle\int H) \xrightarrow{\iota} E\Sigma_n \underset{\Sigma_n}{\times} (BH)^n \xrightarrow{1 \underset{\Sigma_n}{\times} (\alpha_H(X))^n} E\Sigma_n \underset{\Sigma_n}{\times} (QS^0)^n \xrightarrow{\xi} QS^0.$$

Note: It's easy to see

$$\mathfrak{P}^n_{\oplus} = \mathrm{Ind}^{\oplus \Sigma_n \int H}_{H\times\Sigma_{n-1}\int H} \circ p^* \text{ and } \mathfrak{P}^n_{\otimes} = \mathrm{Ind}^{\otimes \Sigma_n \int H}_{H\times\Sigma_{n-1}\int H} \circ p^*,$$

where p^* is induced by the first projection p: $H\times\Sigma_{n-1}\int H \to H$ and $H\times\Sigma_{n-1}\int H$ is regarded as a subgroup of $\Sigma_n\int H$, consisting of those $(\sigma;\ h_1,\ldots,h_n)\in\Sigma_n\int H$ such that $\sigma(1)=1$. And it's also easy to see

$$\mathrm{Ind}^{\oplus G}_{H} = \Psi^* \circ \mathfrak{P}^n_{\oplus} \text{ and } \mathrm{Ind}^{\otimes G}_{H} = \Psi^* \circ \mathfrak{P}^n_{\otimes}.$$

Therefore Lemma 1.2 and Lemma 1.3 are essentially the same statement.

The proof of these are given in the (equivariant) stable category [11], with which both the Burnside ring and QS^0 (especially its multiplicative structure) have intimate relationship. But most of these results, except the multiplicative case of Lemma 1.2 and Lemma 1.3, are well known. So we just give a proof of the multiplicative case of Lemma 1.3 (for complete proofs of other cases, see e.g. [27]).

To define α_G, we first recall the tom Dieck's isomorphism [6][11,V.2.11.];

$$\chi\colon A(G) \simeq \pi^0_G(S),$$

which sends a compact G-manifold F to the composition of

$$\tau(F)\colon S^V \xrightarrow{t} T\nu \xrightarrow{e} T(\nu\oplus\tau) \cong S^V \wedge F_+$$

and the collapse map (Throughout this section any collapse map is denoted by c)

$$1\wedge c\colon S^V \wedge F_+ \longrightarrow S^V \wedge S^0 \cong S^V.$$

Here F is smoothly embedded in V with normal bundle ν, and t is the Pontryagin-Thom map. And τ is the tangent bundle of F, and e is induced by the inclusion of ν in $\nu\oplus\tau$. (See [6][11, IV.2.1, IV.2.3, and V.1.1.])

From this the required ring homomomorphism

$$\alpha_G\colon A(G) \longrightarrow \pi^0_s(BG_+),$$

is defined as follows: Consider the following

$$\begin{array}{ccc} [\Sigma^\infty EG_+,\ \Sigma^\infty S^0]_G^{\mathcal{U}^G} & \xrightarrow{\ i_*\ } & [\Sigma^\infty EG_+,\ \Sigma^\infty S^0]_G^{\mathcal{U}} \\ \downarrow /G & & \\ [\Sigma^\infty BG_+,\ \Sigma^\infty S^0], & & \end{array}$$

where $[\ ,\ ]_G^{\mathcal{U}}$ (in [11] this is simply denoted by $[\ ,\ ]_G$) stands for the set of morphisms in the stable category $\bar{h}G\mathcal{S}\mathcal{A}$ [11, I.5.] for some indexing set $\mathcal{A}$ of a (complete) G-universe $\mathcal{U}$ [11, I.2.], which is independent of $\mathcal{A}$ [11, I.2.4 and I.2.5.], and i: $\mathcal{U}^G \to \mathcal{U}$ is the inclusion of the G-fixed universe $\mathcal{U}^G$. Then the change of universe functor i_* is an isomorphism by [11, II.2.8.]. In fact the down arrow [11, I.3.7.(i)] is also an isomorphism by [11, II.4.5.], but we don't need this. Now for any compact G-manifold F, using $/G\circ i_*^{-1}$, regard the element

$$\chi(F)\circ\Sigma^\infty c = \chi(F)\wedge\Sigma^\infty c \in [\Sigma^\infty EG_+,\ \Sigma^\infty S^0]_G^{\mathcal{U}}$$

as an element of $[\Sigma^\infty BG_+,\ \Sigma^\infty S^0]$: That's the definition of $\alpha_G(F)$.

When G is a compact Lie group and $F{=}G/H$ for some subgroup H with $|G/H|{=}n$, using the fact

$$\pi_s^0(BG_+) \cong \varprojlim_i \pi_s^0(BG_+^{(i)}),$$

and the equivalence of the transfers of Kahn-Priddy [10], Becker-Gottlieb [3], and Lewis-May-Steinberger [11] when the base space is finite (see [22], [4], and [11, IV.3.3.(iii)]) we find the adjoint of the above composite becomes

$$BG \xrightarrow{B\Psi} B(\Sigma_n\int H) \simeq E\Sigma_n\underset{\Sigma_n}{\times}(BH)^n \xrightarrow{1\underset{\Sigma_n}{\times}(c)^n} E\Sigma_n\underset{\Sigma_n}{\times}([1])^n \xrightarrow{D\text{-}L} Q_nS^0,$$

where Ψ is the usual homomorphism [8][10], D-L is the Dyer-Lashof map, and this composite is nothing but the simple composite

$$BG \longrightarrow B\Sigma_n \xrightarrow{D\text{-}L} Q_nS^0,$$

where the first map is induced by the homomorphism

$$G \longrightarrow \operatorname{Aut}(G/H) \simeq \Sigma_n.$$

From this property, we can express (for example) the composite

$$B\Sigma_n \xrightarrow{D\text{-}L} Q_nS^0 \xrightarrow{*[k]} Q_{n+k}S^0$$

as $\alpha_{\Sigma_n}(\Sigma_n/\Sigma_{n-1}+k)$. Of course, when $k<0$ this expression of the element of the Burnside ring is just a formal sum (as an element of the Grothendieck ring of finite G-sets). In this case we can re-expresss

$$\Sigma_n/\Sigma_{n-1}+k = \Sigma_n/\Sigma_{n-1} + (-k)\times T \in A(\Sigma_n),$$

where T is a trivial Σ_n-manifold with $\chi(T)=-1$. (Any T like this is -1 in the tom Dieck's Burnside ring and this is the reason why the tom Dieck's Burnside ring is a set of equivalence classes not merely a formal sum and subtraction) Since the most important cases are $k=-n$ (additive case; $Q_0(S^0)$) and $k=1-n$ (multiplicative case; $Q_1(S^0)$), this property turns out to be very useful.

Finally we recall the fact that SG $(=Q_1S^0)$ is an infinite loop space and that $\xi|_{E\Sigma_n \underset{\Sigma_n}{\times} (SG)^n}$ may be thought as the Dyer-Lashof map of SG, is obtained by restricting the multiplicative operad action on QS^0 to a connected space SG and then applying [13, 14.4.]. Now we express the multiplicative structure of QS^0 stably, which is indispensible for our proof of Lemma 1.3.

Key Fact ([11] see also [14][26]) *The multiplicative operad action ξ is given by the adjoint of*

$$\Sigma^\infty\Big(E\Sigma_{n+} \underset{\Sigma_n}{\wedge} (QS^0)^{(n)}\Big) = \Sigma^\infty D_n(QS^0) \simeq D_n(\Sigma^\infty QS^0)$$

$$\xrightarrow{D_n(\epsilon)} D_n(\Sigma^\infty S^0) = \Sigma^\infty D_n(S^0) \simeq \Sigma^\infty B\Sigma_{n+} \xrightarrow{\Sigma^\infty c} \Sigma^\infty S^0,$$

where $(QS^0)^{(n)}$ is the n-fold smash product of QS^0 whose base point is the trivial one in Q_0S^0, ϵ: $\Sigma^\infty QS^0 \longrightarrow \Sigma^\infty S^0$ is the adjoint of the identity map $QS^0 \longrightarrow QS^0$, $D_n(?)$ is the extended power of the spectrum ? or extended power of the based space ? depending upon ? [28][11, VI.], *and the two equivalences follow since D_n and Σ^∞ commutes* [28, I.2.3.] [11, VI.5.3.].

Proof. We only have to show the homotopy commutativity of the following diagram:

$$\begin{array}{ccccccc}
\Sigma^\infty\Big(E\Sigma_{n+}\underset{\Sigma_n}{\wedge}(QS^0)^{(n)}\Big)\simeq D_n(\Sigma^\infty QS^0) & & \xrightarrow{D_n(\epsilon)} & & D_n(\Sigma^\infty S^0)\simeq\Sigma^\infty B\Sigma_{n+} \\
\Sigma^\infty\xi(QS^0)\downarrow \quad (1) & & \xi(\Sigma^\infty QS^0)\downarrow \quad (2) & & \xi(\Sigma^\infty S^0)\downarrow \quad (3) & & \downarrow\Sigma^\infty c \\
\Sigma^\infty QS^0 & = & \Sigma^\infty QS^0 & \xrightarrow{\epsilon} & \Sigma^\infty S^0 & = & \Sigma^\infty S^0
\end{array}$$

Now the argument goes as follows: Since S^0 is an E_∞ ring space [14, IV.1.5.], $\Sigma^\infty S^0$ is an E_∞ ring spectrum [14, IV.1.7.], and $QS^0=(\Sigma^\infty S^0)_0$ is an E_∞ ring space [14, IV.1.9.]. (Warning: in [14] the notation Q_∞ is used for Σ^∞, see [15, p.55 A.2.(3)].) Then the adjoint ϵ: $\Sigma^\infty QS^0 \longrightarrow \Sigma^\infty S^0$ of the identity map $QS^0 \longrightarrow QS^0$ of E_∞ ring space QS^0 is a map of E_∞ ring spectra [14, IV.1.8.]. Then the commutativity of (2) is just the naturality of the structure maps of E_∞ ring spectra [11, VII.2.4.]. (Warning: [14, IV.1.1.(d)] wasn't good enough to construct the structure map and was replaced by [11, VII.2.4.(iv)], but there's no effect on the above argument.) From [14, IV.1.5.&1.6.][11, VI.5.3.&VII.2.4.] the commutativity of (1) (resp. (3)) follows since the E_∞ ring spectrum structure of $\Sigma^\infty QS^0$ (resp. $\Sigma^\infty S^0$) comes from the E_∞ ring space structure of QS^0 (resp. S^0). □

Proof of the multiplicative case of Lemma 1.3. Let $\mathcal{A}$ be an index set in the H-universe $\mathcal{U}$. Then we form the $(\Sigma_n\int H)$-universe $\mathcal{P}^n_\oplus\mathcal{U}$, which is as a space equivalent to $\mathcal{U}^n$ with $\Sigma_n\int H$ action

$$(\sigma;\, h_1,\ldots,h_n)(u_1,\ldots,u_n) = (h_1 u_{\sigma^{-1}(1)},\ldots,h_n u_{\sigma^{-1}(n)}),$$

and let $\mathcal{P}^n_\oplus\mathcal{A}=\{V^n:=\mathcal{P}^n_\oplus V|\ V\in\mathcal{A}\}$ be the resulting index set in the $(\Sigma_n\int H)$-universe $\mathcal{P}^n_\oplus\mathcal{U}$. We have to give two comments here. Firstly the reason, why $\oplus$ is used as subscripts instead of $\otimes$, is in the representation theory the direct sum (which happens to be the product set-theoretically) gives the additive structure (of course the tensor product gives the multiplicative structure). Secondly we have to check that $\mathcal{P}^n_\oplus\mathcal{U}$ is really the $(\Sigma_n\int H)$-universe, when $\mathcal{U}$ is a H-universe. Recall that a universe is defined by the property of containing a trivial representation and each of its finite subrepresentations infinitely often [11, I.2.]. Now note that

$$\mathcal{P}^n_{\oplus} = \mathrm{Ind}^{\oplus \Sigma_n \int H}_{H \times \Sigma_{n-1} \int H} \circ p^*,$$

where $\mathrm{Ind}^{\oplus}$ is the induced representation and $p\colon H \times \Sigma_{n-1} \int H \to H$ is the projection onto the first factor. Then we only have to check that $\mathrm{Ind}^{\oplus}$ sends a universe to a universe; which is very easy to see.

Now we define the extended smash power product:

$$\mathcal{P}^n_{\otimes}\colon H\mathcal{S}\mathcal{A} \longrightarrow (\Sigma_n \int H)\mathcal{S}(\mathcal{P}^n_{\oplus}\mathcal{A})$$

As is usual the case of the convenient setting of coordinated-free spectra, we first define this at the prespectra level and apply L [11, I.2.2.]. For an H-prespectrum $D \in H\mathcal{P}\mathcal{A}$, we define the $(\Sigma_n \int H)$-prespectrum $\mathcal{P}^n_{\otimes}(D) \in (\Sigma_n \int H)\mathcal{P}(\mathcal{P}^n_{\oplus}\mathcal{A})$ by

$\mathcal{P}^n_{\otimes}(D)(\mathcal{P}^n_{\oplus} V) = DV \wedge DV \wedge \ldots \wedge DV$ (n-fold smash product of DV with itself) with the usual $(\Sigma_n \int H)$-action. Given the structure map

$$\sigma\colon \Sigma^{W-V} DV \longrightarrow DW,$$

where $W-V$ is the orthogonal complement for $V \subseteq W$ in $\mathcal{A}$ [11, I.2.1.], we assign the structure map to $\mathcal{P}^n_{\otimes}(D)$ by

$$\Sigma^{(\mathcal{P}^n_{\oplus} W)-(\mathcal{P}^n_{\oplus} V)} \mathcal{P}^n_{\otimes}(D)(\mathcal{P}^n_{\oplus} V) \cong (\Sigma^{W-V} \wedge \Sigma^{W-V} \wedge \ldots \wedge \Sigma^{W-V})(DV \wedge DV \wedge \ldots \wedge DV)$$

$$\cong (\Sigma^{W-V} DV) \wedge \ldots \wedge (\Sigma^{W-V} DV) \xrightarrow{\sigma \wedge \ldots \wedge \sigma} DW \wedge \ldots \wedge DW = \mathcal{P}^n_{\otimes}(D)(\mathcal{P}^n_{\oplus} W).$$

We must check the commutativity of [11, I.2.1.(ii)], which follows from the commutativity of the following diagram, where $a = W-Z$ and $b = V-W$ and the arrows are either some shuffle maps or (possibly some suspension of) the power smash product of the structure map of D:

$$\begin{array}{ccccc}
(\Sigma^a\wedge..\Sigma^a)(\Sigma^b\wedge..\Sigma^b)(DV\wedge..DV) & \to & (\Sigma^a\wedge..\Sigma^a)(\Sigma^b DV\wedge..\Sigma^b DV) & \to & (\Sigma^a\wedge..\Sigma^a)(DW\wedge..\wedge DW) \\
\| & & \downarrow & & \downarrow \\
(\Sigma^a\wedge..\wedge\Sigma^a)(\Sigma^b\wedge..\wedge\Sigma^b)(DV\wedge..\wedge DV) & \to & (\Sigma^a\Sigma^b DV)\wedge..(\Sigma^a\Sigma^b DV) & \to & (\Sigma^a DW)\wedge..\wedge(\Sigma^a DW) \\
\downarrow & & \downarrow\cong & & \downarrow \\
(\Sigma^{a+b}\wedge..\wedge\Sigma^{a+b})(DV\wedge..\wedge DV) & \longrightarrow & (\Sigma^{a+b}DV)\wedge..\wedge(\Sigma^{a+b}DV) & \longrightarrow & DZ\wedge..\wedge DZ,
\end{array}$$

When restricted to $H\times\Sigma_n$-action (recall that $H\times\Sigma_n\subseteq\Sigma_n\int H$ using the diagonal embedding $H\to H^n$) the above construction recovers the construction

$$(?)^{(n)}:\ H\mathcal{S}\mathcal{A}\to(H\times\Sigma_n)\mathcal{S}\mathcal{A}^n$$

in [11, VI.5.], and in particular when H is trivial these constructions are exactly the same. (For Lewis-May-Steinberger, the interest was to construct the equivariant extended power of the spectra and for that purpose considerations in $(H\times\Sigma_n)\mathcal{S}\mathcal{A}$ was sufficient) Concerning this we note one remark; given H-universe $\mathcal{U}$ and its indexing set $\mathcal{A}$

$$\mathcal{P}^n_{\oplus}(\mathcal{U}^H)=(\mathcal{P}^n_{\oplus}\mathcal{U})^{H^n},\quad \mathcal{P}^n_{\oplus}(\mathcal{A}^H)=(\mathcal{P}^n_{\oplus}\mathcal{A})^{H^n},$$

where $\mathcal{A}^H=\{V^H\mid V\in\mathcal{A}\}$, and the following diagram commutes:

$$\begin{array}{ccc}
H\mathcal{S}(\mathcal{A}^H) & \xrightarrow{\mathcal{P}^n_{\otimes}} & (\Sigma_n\int H)\mathcal{S}(\mathcal{P}^n_{\oplus}(\mathcal{A}^H)) \\
\downarrow/H & & \downarrow/H^n \\
\mathcal{S}(\mathcal{A}^H) & \xrightarrow{(?)^n} & \Sigma_n\mathcal{S}(\mathcal{P}^n_{\oplus}(\mathcal{A}^H))
\end{array}$$

Now we are in a position to show the commutativity of the following diagram

$$
\begin{array}{ccccc}
A(H) & \xrightarrow{\qquad\mathfrak{P}^n_{\otimes}\qquad} & & & A(\Sigma_n\int H) \\
\simeq\downarrow\chi & & (1) & & \simeq\downarrow\chi \\
\{S^0,S^0\}^{\mathfrak{U}}_H & \xrightarrow{\qquad\mathfrak{P}^n_{\otimes}\qquad} & & & \{S^0,S^0\}^{\mathfrak{P}(\mathfrak{U})}_{\Sigma_n\int H} \\
\downarrow\Sigma^\infty c^* & & (2) & & \downarrow\Sigma^\infty c^* \\
\{EH_+,S^0\}^{\mathfrak{U}}_H & \xrightarrow{P} & \{E\Sigma_{n+}\wedge(EH_+)^n,S^0\}^{\mathfrak{P}(\mathfrak{U})}_{\Sigma_n\int H} & = & \{E\Sigma_{n+}\wedge(EH_+)^n,S^0\}^{\mathfrak{P}(\mathfrak{U})}_{\Sigma_n\int H} \\
\uparrow\simeq & (3) & \uparrow\simeq & (4) & \uparrow\simeq \\
\{EH_+,S^0\}^{\mathfrak{U}^H}_H & \xrightarrow{P} & \{E\Sigma_{n+}\wedge(EH_+)^n,S^0\}^{\mathfrak{P}(\mathfrak{U}^H)}_{\Sigma_n\int H} & \simeq & \{E\Sigma_{n+}\wedge(EH_+)^n,S^0\}^{\mathfrak{P}(\mathfrak{U}^H)^{\Sigma_n}}_{\Sigma_n\int H} \\
\downarrow/H & (5) & \downarrow/H^n & (6) & \downarrow/\Sigma_n\int H \\
\{BH_+,S^0\}^{\mathfrak{U}^H} & \xrightarrow{P} & \{E\Sigma_{n+}\wedge(BH_+)^n,S^0\}^{\mathfrak{P}(\mathfrak{U}^H)}_{\Sigma_n} & \xrightarrow{Q} & \{E\Sigma_{n+}\wedge_{\Sigma_n}(BH_+)^n,S^0\}^{\mathfrak{P}(\mathfrak{U}^H)^{\Sigma_n}} \\
\| & (7) & \uparrow\Sigma^\infty c_* & (8) & \uparrow\Sigma^\infty c_* \\
\{BH_+,S^0\}^{\mathfrak{U}^H} & \xrightarrow{P'} & \{E\Sigma_{n+}\wedge(BH_+)^n,E\Sigma_{n+}\}^{\mathfrak{P}(\mathfrak{U}^H)}_{\Sigma_n} & \xrightarrow{Q} & \{E\Sigma_{n+}\wedge(BH_+)^n,B\Sigma_{n+}\}^{\mathfrak{P}(\mathfrak{U}^H)^{\Sigma_n}}
\end{array}
$$

where we abbreviated $\{\Sigma^\infty X,\Sigma^\infty Y\}$ as $\{X,Y\}$, c: $(?)_+\to S^0$ is the collapse map, and the isomorphic property of the three up arrows is the consequence of the change of universe theorem [11, II.2.8.]. Furthermore $P=(\Sigma^\infty c_*)\circ(1\wedge\mathfrak{P}^n_\otimes)$, $P'=1\wedge\mathfrak{P}^n_\otimes$ and $Q=(/\Sigma_n)\circ(i_*^{-1})$, where i: $\mathfrak{P}(\mathfrak{U}^H)^{\Sigma_n}\to\mathfrak{P}(\mathfrak{U}^H)$ induces the change of universe isomorphism i_* [11, II.2.8.]. Then, (2) and (3) respectively commute because of the naturality of $\mathfrak{P}^n_\otimes$ with respect to the spectra and the change of universe respectively. The commutativity of (4) is trivial. The commutativity of (5) follows from the comment just before the diagram. The commutativity of (6) is nothing but $(/\Sigma_n)\circ(/H^n)=/\Sigma_n\int H$. (7) commutes from the definitions. The commutativity of (8) follows from the naturality of the change of universe and $/\Sigma_n$. Now to show the commutativity of (1), we pick up two compact G-manifolds M. From the definition [11, IV.2.3.], $\chi(\mathfrak{P}^n_\otimes(M))$ is obtained (at the space level) as the composite

$$S^V \xrightarrow{\tau(M^n)} S^V\wedge(M^n)_+ \xrightarrow{1\wedge c} S^V\wedge S^0\simeq S^V,$$

where $\tau(M^n)$ is given by the Pontryagin-Thom construction using a $\Sigma_n\int H$-embedding of M^n $(=M\times\ldots\times M)$ in V. The point is that we could have begun from the H-embedding $M\subseteq W$, taken the n-fold power of this to obtain $\Sigma_n\int H$-embedding $M^n\subseteq W^n=W\oplus\ldots\oplus W=\mathfrak{P}^n_\oplus W$, and put $V=\mathfrak{P}^n_\oplus W$. Then we see, from our definition of $\mathcal{P}^n_{\otimes}$, (in the stable category)

$$\tau(M^n)=\mathcal{P}^n_{\otimes}(\tau(M)) \colon \Sigma^{\infty}S^0=\mathcal{P}^n_{\otimes}(\Sigma^{\infty}S^0) \longrightarrow \mathcal{P}^n_{\otimes}(\Sigma^{\infty}M_+) \simeq \Sigma^{\infty}(M^n)_+,$$

which immediately implies the desired result $\chi(\mathcal{P}^n_{\otimes}(M))=\mathcal{P}^n_{\otimes}(\chi(M))$.

Now we express the stable adjoint of $\alpha_H(M) \colon BH_+ \to QS^0$ as

$$\Sigma^{\infty}BH_+ \xrightarrow{\Sigma^{\infty}\alpha_H(M)} \Sigma^{\infty}QS^0 \xrightarrow{\ \epsilon\ } \Sigma^{\infty}S^0,$$

where $\epsilon \colon \Sigma^{\infty}QS^0 \longrightarrow \Sigma^{\infty}S^0$ is the stable adjoint of the identity map $QS^0 \longrightarrow QS^0$. Then from the commutativity of the above diagram, we see that the stable adjoint of $\alpha_{\Sigma_n \int H}(\mathcal{P}^n_{\otimes}(X))$ is the composite

$$\Sigma^{\infty}E\Sigma_n \underset{\Sigma_n}{\times} (BH)^n{}_+ \simeq \Sigma^{\infty}E\Sigma_{n+} \underset{\Sigma_n}{\wedge} (BH_+)^{(n)} \xrightarrow{\Sigma^{\infty}(1_+ \wedge_{\Sigma_n} (\alpha_H(M))^{(n)})} \Sigma^{\infty}E\Sigma_{n+} \underset{\Sigma_n}{\wedge} (QS^0)^{(n)}$$

$$\simeq \Sigma^{\infty}E\Sigma_{n+} \underset{\Sigma_n}{\wedge} (\Sigma^{\infty}QS^0)^{(n)} \xrightarrow{\Sigma^{\infty}(1_+ \wedge_{\Sigma_n} (\epsilon)^{(n)})} \Sigma^{\infty}E\Sigma_{n+} \underset{\Sigma_n}{\wedge} (\Sigma^{\infty}S^0)^{(n)} \simeq \Sigma^{\infty}B\Sigma_{n+} \xrightarrow{\Sigma^{\infty}c} \Sigma^{\infty}S^0.$$

Note that the composition in the first line is essentially the stablization of the unstable map

$$E\Sigma_n \underset{\Sigma_n}{\times} (BH)^n \xrightarrow{1 \times_{\Sigma_n} (\alpha_H(M))^n} E\Sigma_n \underset{\Sigma_n}{\times} (QS^0)^n,$$

and the composition in the second line is the stable adjoint of the multiplicative operad action

$$\xi \colon E\Sigma_n \underset{\Sigma_n}{\times} (QS^0)^n \longrightarrow QS^0$$

by the Key Fact. Therefore the conclusion follows. □

§2. CALCULATIONS IN THE BURNSIDE RING

In this section we give some calculational results of the Burnside ring. This consists of essentially two different parts; one is related to the deviation from the additivity of the multiplicative transfer (in homology operation this is the mixed Cartan formula) and the other is related to the existence of p-adic exponential and p-adic logarithm. (The latter is the place where the essential difference between 2 and odd primes occurs) Combining these two results, examples which appear in the calculation of the homology operation are studied. From the property of the homology operations, many of the calculations are reduced to the case when the group is abelian (especially elementary abelian p-group), which makes the calculation tractable.

In the rest of the paper we fix the following specific chain of subgroups of $\Sigma_n \int H$ (, where H is some group) and Σ_{p^k}:

$$\Sigma_n \int H \supseteq \Sigma_i \int H \times \Sigma_{n-i} \int H \supseteq H$$

$$\Sigma_n \int H \supseteq (\mathbb{Z}/n) \int H \supseteq (\mathbb{Z}/n) \times H \supseteq H,$$

where

$$\Sigma_i \int H \times \Sigma_{n-i} \int H = \{(\sigma;\, h_1, \cdots, h_n) \mid \sigma \in \Sigma_i \times \Sigma_{n-i} \text{ and } h_k \in H \text{ for } 1 \leq k \leq n\} \subseteq \Sigma_n \int H\,,$$

$$H = \{(e;\, h, \cdots, h) \mid e = \text{unit of} \Sigma_n, \text{ and } h \in H\} \subseteq \Sigma_n \int H\,,$$

and $\mathbb{Z}/n$ is regarded as the cyclic subgroup of Σ_n generated by the permutation

$$(1,2,\cdots,n) = \begin{pmatrix} 1 & 2 & \cdots & n \\ 2 & 3 & \cdots & 1 \end{pmatrix};$$

$$(\mathbb{Z}/n) \int H = \{(\tau;\, h_1, \cdots, h_n) \mid \tau \in \mathbb{Z}/n \subseteq \Sigma_n, \text{ and } h_k \in H \text{ for } 1 \leq k \leq n\} \subseteq \Sigma_n \int H\,,$$

$$(\mathbb{Z}/n) \times H = \{(\tau;\, h, \cdots, h) \mid \tau \in \mathbb{Z}/n \subseteq \Sigma_n, \text{ and } h \in H\} \subseteq \Sigma_n \int H\,.$$

Quite often $n=p$, a prime number, and $H = (\mathbb{Z}/p)^r$ for some r. Later in §3 (beginning from Prop. 3.2), we consider the very important standard inclusion

$$(\mathbb{Z}/p)^k \subseteq \mathcal{I}(p^k,p) : \text{the } k\text{-fold iterated wreath product } \mathbb{Z}/p\int \cdots \int\mathbb{Z}/p .$$

This is obtained by successively applying one of the above-mentioned inclusion

$$(\mathbb{Z}/n)\times H \subseteq (\mathbb{Z}/n)\int H ;$$

$$(\mathbb{Z}/p)^k = (\mathbb{Z}/p)\times(\mathbb{Z}/p)^{k-1} \subseteq (\mathbb{Z}/p)\int(\mathbb{Z}/p)^{k-1} \subseteq (\mathbb{Z}/p)\int\mathcal{I}(p^{k-1},p) \subseteq \mathcal{I}(p^k,p) .$$

Of course, using the canonical inclusion

$$\Sigma_a\int\Sigma_b \subseteq \Sigma_{ab}$$

and the above-mentioned

$$\mathbb{Z}/p \subseteq \Sigma_p,$$

we get the usual embedding

$$\mathcal{I}(p^k,p) \to \Sigma_{p^k} ,$$

which makes $\mathcal{I}(p^k,p)$ as a p-Sylow subgroup of Σ_{p^k}.

Now we present some results related to the multiplicative transfer.

Proposition 2.1. *For any compact Lie group H and any compact H-manifolds M and N,*

$$(M\coprod N)^n = M^n \coprod N^n \coprod_{i=1}^{n-1} \frac{(\Sigma_n\int H) \times (M^i\times N^{n-i})}{(\Sigma_i\int H\times\Sigma_{n-i}\int H)}$$

as compact $\Sigma_n\int H$-manifolds. Here (for all $0\le i\le n$) we are regarding $M^i\times N^{n-i}=\mathcal{P}^i_\otimes(M)\times\mathcal{P}^{n-i}_\otimes(N)$ as $(\Sigma_i\int H\times\Sigma_{n-i}\int H)$-manifold using $\Sigma_i\int H$ (resp. $\Sigma_{n-i}\int H$) action on M^i (resp. N^{n-i}), given in Lemma 1.3 .

Therefore, if we abuse the notation and regard M, N *as elements of* $A(H)$,

$$\mathcal{P}^n_{\otimes}(M+N) = \mathcal{P}^n_{\otimes}(M) + \mathcal{P}^n_{\otimes}(N) + \sum_{i=1}^{n-1} \operatorname{Ind}^{\oplus}{}^{\Sigma_n\int H}_{\Sigma_i\int H\times\Sigma_{n-i}\int H} (\mathcal{P}^i_{\otimes}(M)\times\mathcal{P}^{n-i}_{\otimes}(N))$$

in $A(\Sigma_n\int H)$.

Proof. Let S_i $(0\le i\le n)$ be the subset of $(M\coprod N)^n$ such that exactly i members of n coordinates belong to M. Then it's easy to see that

$$S_i = (\Sigma_n\int H) \underset{(\Sigma_i\int H\times\Sigma_{n-i}\int H)}{\times} (M^i\times N^{n-i}),$$

which proves the proposition. □

Lemma 2.2. *Suppose* H *is an abelian group,* p *is a prime number, and* $1\le i\le p-1$, *then*

$$\operatorname{Res}^{\Sigma_p\int H}_{(\mathbb{Z}/p)\times H} \operatorname{Ind}^{\oplus}{}^{\Sigma_p\int H}_{\Sigma_i\int H\times\Sigma_{p-i}\int H} = \frac{1}{p}\binom{p}{i} \operatorname{Ind}^{\oplus}{}^{(\mathbb{Z}/p)\times H}_{H} \operatorname{Res}^{\Sigma_i\int H\times\Sigma_{p-i}\int H}_{H}.$$

Proof. We claim that, under these condition,

$$C_{g^{-1}} \operatorname{Ind}^{\oplus}{}^{((\mathbb{Z}/p)\times H)^g}_{(\Sigma_i\int H\times\Sigma_{p-i}\int H)\cap((\mathbb{Z}/p)\times H)^g} \operatorname{Res}^{\Sigma_i\int H\times\Sigma_{p-i}\int H}_{(\Sigma_i\int H\times\Sigma_{p-i}\int H)\cap((\mathbb{Z}/p)\times H)^g}$$

$$= \operatorname{Ind}^{\oplus}{}^{(\mathbb{Z}/p)\times H}_{H} \operatorname{Res}^{\Sigma_i\int H\times\Sigma_{p-i}\int H}_{H},$$

for any $g\in\Sigma_p\int H$, where $(?)^g = g^{-1}(?)\,g$ and $C_{g^{-1}}$ is the conjugation by g^{-1}, i.e. $C_{g^{-1}} : x\mapsto gxg^{-1}$. This would prove the lemma by the double coset formula (e.g. [6][11]). In fact, from the hypothesis,

$$(\Sigma_i\int H\times\Sigma_{p-i}\int H)\cap((\mathbb{Z}/p)\times H)^g = H \text{ (this is the diagonal subgroup of } H^p)$$

and H is the center of $\Sigma_p \int H$ except the trivial case: $p=2$ and $H=\{e\}$. Therefore

$$C_{g^{-1}} \operatorname{Ind}^{\oplus\,((\mathbb{Z}/p)\times H)^g}_{(\Sigma_i\int H\times\Sigma_{p-i}\int H)\cap((\mathbb{Z}/p)\times H)^g} \operatorname{Res}^{\Sigma_i\int H\times\Sigma_{p-i}\int H}_{(\Sigma_i\int H\times\Sigma_{p-i}\int H)\cap((\mathbb{Z}/p)\times H)^g}$$

$$= C_{g^{-1}} \operatorname{Ind}^{\oplus((\mathbb{Z}/p)\times H)^g}_{H} \operatorname{Res}^{\Sigma_i\int H\times\Sigma_{p-i}\int H}_{H}$$

$$= \operatorname{Ind}^{\oplus(\mathbb{Z}/p)\times H}_{H^{g-1}} C_{g^{-1}} \operatorname{Res}^{\Sigma_i\int H\times\Sigma_{p-i}\int H}_{H}$$

$$= \operatorname{Ind}^{\oplus(\mathbb{Z}/p)\times H}_{H} \operatorname{Res}^{\Sigma_i\int H\times\Sigma_{p-i}\int H}_{H},$$

as desired. □

Corollary 2.3. *If M is a compact free H-manifold, then*

$$\coprod_{i=1}^{p-1} (\Sigma_p\int H) \underset{(\Sigma_i\int H\times\Sigma_{p-i}\int H)}{\times} (M^i\times N^{p-i})$$

is a compact free $(\mathbb{Z}/p)\times H$ manifold. (Note that the concept "free" is equally meaningful at the manifold level and at thc Burnsidc ring lcvcl)

Proof. When M is a compact free H-manifold, so is $M^i\times N^{p-i}$. Thus the claim follows from the above lemma and the fact that $\operatorname{Ind}^{\oplus}$ sends "free" to "free" (both at the manifold level and at the Burnside ring level). □

Corollary 2.4. *Let* $X=(\mathbb{Z}/p)^r\in A((\mathbb{Z}/p)^r)$ *and* $Y=1-p^r\in A((\mathbb{Z}/p)^r)$, *then*

$$\operatorname{Res}^{\Sigma_p\int(\mathbb{Z}/p)^r}_{(\mathbb{Z}/p)\times(\mathbb{Z}/p)^r}\Big(\sum_{i=1}^{p-1}\operatorname{Ind}^{\oplus}{}^{\Sigma_p\int(\mathbb{Z}/p)^r}_{\Sigma_i\int(\mathbb{Z}/p)^r\times\Sigma_{p-i}\int(\mathbb{Z}/p)^r}(\mathcal{P}^i_{\otimes}(X)\times\mathcal{P}^{p-i}_{\otimes}(Y))\Big)$$

$$= k\,((\mathbb{Z}/p)\times(\mathbb{Z}/p)^r)\in A((\mathbb{Z}/p)\times(\mathbb{Z}/p)^r),$$

where $k=\dfrac{1-(p^r)^p-(1-p^r)^p}{p\times p^r}\equiv 1 \pmod{p}$.

Proof. From the above corollary, we see that the questioned element is a some multiple of $(\mathbb{Z}/p)\times(\mathbb{Z}/p)^r\in A((\mathbb{Z}/p)\times(\mathbb{Z}/p)^r)$. Now the claim follows from

$$\frac{\epsilon(\mathcal{P}^p_{\otimes}(X+Y)-\mathcal{P}^p_{\otimes}(X)-\mathcal{P}^p_{\otimes}(Y))}{\epsilon((\mathbb{Z}/p)\times(\mathbb{Z}/p)^r)}=\frac{1-(p^r)^p-(1-p^r)^p}{p\times p^r}\equiv 1 \pmod{p},$$

where ϵ is the augmentation. □

Lemma 2.5. *We suppose* $r\geq 2$ *when* p *is odd, and* $r\geq 1$ *when* $p=2$. *Then*

$$\operatorname{Res}^{\Sigma_p\int(\mathbb{Z}/p)^r}_{(\mathbb{Z}/p)\times(\mathbb{Z}/p)^r}\mathcal{P}^p_{\otimes}((\mathbb{Z}/p)^r)$$

$$= l\,((\mathbb{Z}/p)\times(\mathbb{Z}/p)^r)+\sum_{\substack{|H_i|=p\\ H_i\subset(\mathbb{Z}/p)\times(\mathbb{Z}/p)^r}} a_i\,((\mathbb{Z}/p)\times(\mathbb{Z}/p)^r)/H_i$$

for some $a_i\in\mathbb{Z}$ *and* $l=p^{r-1}((p^r)^{p-2}-1)\equiv 0 \pmod{p}$.

Proof. For simplicity, we put $H=(\mathbb{Z}/p)^r$. First we claim

$$\#|((\mathbb{Z}/p)\times H)_{(x_1,x_2,\ldots,x_p)}|=1, \text{ or } p, \text{ for any } (x_1,x_2,\ldots,x_p)\in H^p,$$

where $((\mathbb{Z}/p)\times H)_{(x_1,x_2,\ldots,x_p)}$ is the isotropy subgroup of $(\mathbb{Z}/p)\times H$ at $(x_1,x_2,\ldots,x_p)$

$\in H^p$, under $(\mathbb{Z}/p)\times H \subseteq \Sigma_p \int H$ action on H^p induced (as was given in Lemma 1.3) from the H-left multiplication action on H.

To see this, we note that

$$((\mathbb{Z}/p)\times H)_{(x_1,x_2,\dots,x_p)} \bigcap H = \{e\},$$

since H acts freely on H^p. And this means that the composite

$$((\mathbb{Z}/p)\times H)_{(x_1,x_2,\dots,x_p)} \subseteq (\mathbb{Z}/p)\times H \to ((\mathbb{Z}/p)\times H)/H \simeq \mathbb{Z}/p$$

is injective, which implies the claim. Now we only have to count the number of free $((\mathbb{Z}/p)\times H)$-orbits in H^p. For this, we have

$((\mathbb{Z}/p)\times H)_{(x_1,x_2,\dots,x_p)} \neq \{e\}$

$\Leftrightarrow \exists(\sigma,h)\in((\mathbb{Z}/p)\times H)_{(x_1,x_2,\dots,x_p)}$ such that $\sigma\neq e$

$\Leftrightarrow \exists(\tau,h)\in((\mathbb{Z}/p)\times H)_{(x_1,x_2,\dots,x_p)}$, where $\tau(i)=i+1$ for $1\leq i\leq p-1$ and $\tau(p)=1$

$\Leftrightarrow (x_1,x_2,\dots,x_p) = (hx_p,hx_1,\dots,hx_{p-1})$ for some $h\in H$

$\Leftrightarrow x_i = h^{i-1}x_1,\ 1\leq i\leq p$ for some $h\in H$.

Thus the number of points in H^p with *non*-free $((\mathbb{Z}/p)\times H)$-orbit is $\#|H|\times\#|H| = p^{2r}$, which implies the number of free $((\mathbb{Z}/p)\times H)$-orbits in H^p is

$$\frac{(p^r)^p-p^{2r}}{p\times p^r} = \frac{(p^r)^{p-1}-p^r}{p} = p^{r-1}((p^r)^{p-2}-1) \equiv 0 \text{ (mod. } p)$$

under the hypothesis. This proves the lemma. □

The reader might have noticed that l in Lemma 2.5. is 0 when $p=2$ and non-zero when p is odd. Indeed this is the very fact which has made odd-primary multiplicative homology operations complicated (at least its surface) and caused the problem between Tsuchiya [26] and May [5]. The reader would notice that this isn't the essential difference between 2 and odd prime; he will find out the essential difference in exp_p and log_p.

Lemma 2.6.

$$\mathrm{Res}^{\Sigma_p \int (\mathbb{Z}/p)^r}_{(\mathbb{Z}/p)\times(\mathbb{Z}/p)^r} \mathcal{P}^p_{\otimes}(1-p^r) = (1-p^r) + m\,((\mathbb{Z}/p)\times(\mathbb{Z}/p)^r)/(\mathbb{Z}/p)^r,$$

where $m = \dfrac{(1-p^r)^p-(1-p^r)}{p} \equiv 0 \ (\mathrm{mod.}\ p^{r-1})$.

Proof. Obviously $1-p^r \in A((\mathbb{Z}/p)^r)$ comes from $A(\{e\})$ by the map π^*: $A(\{e\}) \to A((\mathbb{Z}/p)^r)$, where π: $(\mathbb{Z}/p)^r \to \{e\}$ is the trivial homomorphism. Therefore, from the naturality of $\mathcal{P}^p_{\otimes}$, we only have to prove the case $r=0$. For this, we calculate the Euler number of fixed point sets by $\{e\}$ and $\mathbb{Z}/p$: Let S be a compact manifold with $\chi(S) = 1-p^r$, then

$$\chi(S^p) = \chi(S)^p = (1-p^r)^p \text{ and } \chi\left((S^p)^{\mathbb{Z}/p}\right) = \chi(S) = 1-p^r.$$

Now the conclusion follows since

$$\chi((1-p^r)+\frac{(1-p^r)^p-(1-p^r)}{p}(\mathbb{Z}/p)) = (1-p^r)+\frac{(1-p^r)^p-(1-p^r)}{p}\times p = (1-p^r)^p$$

and

$$\chi\left(((1-p^r) + \frac{(1-p^r)^p-(1-p^r)}{p}(\mathbb{Z}/p))^{\mathbb{Z}/p}\right) = (1-p^r) + 0 = 1-p^r. \qquad \square$$

By now, we have become able to express

$$\mathrm{Res}^{\Sigma_p \int (\mathbb{Z}/p)^r}_{(\mathbb{Z}/p)\times(\mathbb{Z}/p)^r} \mathcal{P}^p_{\otimes}((\mathbb{Z}/p)^r + (1-p^r))$$

additively, but as far as the multiplicative property is concerned, we should express this multiplicatively. For this purpose we are going to study p-adic exponential and p-adic logarithm which interrelates additive and multiplicative structures.

Definition. 2.7. Let A be a finite abelian group and let $I(A)\hat{_p}$ be the p-adic completion of the augmentation ideal of the Burnside ring A. (This time we prefer to adopt the original definition of the Burnside ring) So $I(A)\hat{_p}$ is the free $\mathbb{Z}\hat{_p}$ module generated by $(A/H - \#|A/H|)$, where H runs over the all subgroups of A except A. Now we define various $\mathbb{Z}\hat{_p}$ subideals of $I(A)\hat{_p}$ with the prescribed $\mathbb{Z}\hat{_p}$ basis (in fact, they are free $\mathbb{Z}\hat{_p}$ submodules of $I(A)\hat{_p}$) as follows:

$D(A)\hat{_p}$: $\mathbb{Z}\hat{_p}$ subideal of $I(A)\hat{_p}$ with the basis $(A/H - \#|A/H|)$ such that $\#|A/H| \geq p^2$.

$pI(A)\hat{_p}$: $\mathbb{Z}\hat{_p}$ subideal of $I(A)\hat{_p}$ with the basis $p(A/H - \#|A/H|)$.

Then we set

$$J = \begin{cases} I(A)\hat{_p} & \text{if } p = \text{odd} \\ D(A)\hat{_p} + pI(A)\hat{_p} & \text{if } p = 2 \end{cases}$$

Now we call a sub $\mathbb{Z}\hat{_p}$ module $S \subseteq J$ an <u>*exponentiable*</u> if S has a basis of the form

$$p^{k_i}(A/H_i - \#|A/H_i|), \quad (k_i \geq 0)$$

and satisfies

$$xy \in pS \text{ for any } x, y \in S.$$

Example 2.8. The following are examples of exponentiable modules:

(i) $S = pI(A)\hat{_p}$.

(ii) S : $\mathbb{Z}\hat{_p}$ subideal of $I(A)\hat{_p}$, when p=odd (resp. $\mathbb{Z}\hat{_p}$ subideal of $D(A)\hat{_p}$, when $p=2$) with the basis $(A/H - \#|A/H|)$ such that $\#|A| \geq p(\#|H|)^2$.

(iii) $S + pI(A)\hat{_p}$, where S is an exponentiable ideal.

Theorem 2.9. *Let A be an abelian p-group, then*

(i) *We can define the homomorphism* exp_p: $J \to 1 + J$ *in the usual way;*

$$exp_p(x) = 1+ \sum_{n=1}^{\infty} \frac{x^n}{n!} .$$

(ii) We can define the homomorphism log_p: $1+I(A)\hat{_p} \to I(A)\hat{_p}$ *in the usual way;*

$$log_p(1+x) = \sum_{n=1}^{\infty} (-1)^{n+1} \frac{x^n}{n} .$$

(iii) Suppose $S \subseteq J$ *is a multiplicatively closed* $\mathbb{Z}\hat{_p}$ *submodule with an additive basis of the form*

$$p^{k_i}(A/H_i - \#|A/H_i|), \ (k_i \geq 0) \qquad (e.g.\ S = J,\ pJ,\ D(A)\hat{_p},..).$$

Then exp_p *"preserves"* S: $exp_p(S) \subseteq 1+S$.

(iv) Suppose $S \subseteq I(A)\hat{_p}$ *is a multiplicatively closed* $\mathbb{Z}\hat{_p}$ *submodule with an additive basis of the form*

$$p^{k_i}(A/H_i - \#|A/H_i|), \ (k_i \geq 0) \qquad (e.g.\ S = J,\ pJ,\ D(A)\hat{_p},..).$$

Then log_p *"preserves"* S: $log_p(1+S) \subseteq S$.

(v) $$exp_p \circ (log_p|_{1+J}) = 1_{1+J} \text{ , and } log_p \circ exp_p = 1_J .$$

(vi) Suppose x *is contained in an exponentiable module* S, *then*

$$exp_p(x) \in 1+x+pS = (1+x)(1+pS) \text{ and } log_p(x) \in x+pS.$$

Proof of (i). For any z: an element of either $\mathbb{Z}\hat{_p}$ or $I(A)\hat{_p}$, we put

$$\nu_p(z) = \{\text{the largest integer } n \text{ such that } p^n \text{ divides } z\}.$$

To see that exp_p is well-defined, we want to evaluate

$$\nu_p\left(\frac{\{p^k(A/H - \#|A/H|)\}^n}{n!}\right).$$

(We note we may assume $\nu_p(\#|A/H|) \geq 1$, since otherwise $A/H - \#|A/H| = 0$.) Now it is well-known (and can be checked easily *e.g.* by the induction on n) that $\nu_p(n!) = \frac{n-\alpha(n)}{p-1}$, where $\alpha(n)$ is the sum of p-adic expansion of n. On the other hand, since A is abelian,

$$(A/H - \#|A/H|)^n = (-\#|A/H|)^{n-1}(A/H - \#|A/H|),$$

where we used the fact that

$$(A/K)\times(A/L) = \frac{\#|A|\times\#|(K\cap L)|}{\#|K|\times\#|L|}\,A/(K\cap L)$$

for any subgroups K, L of abelian A. Therefore

$$\nu_p\left(\frac{\{p^k(A/H - \#|A/H|)\}^n}{n!}\right)$$

$$= nk + (n-1)\nu_p(\#|A/H|) - \frac{n-\alpha(n)}{p-1}$$

$$= \frac{nk(p-1) + (n-1)(p-1)\nu_p(\#|A/H|) - n + \alpha(n)}{p-1}$$

$$= \frac{(n-1)\Big\{(p-1)(k+\nu_p(\#|A/H|))-1\Big\}+k(p-1)+\alpha(n)-1}{p-1},$$

which is $\geq k$ since $k\geq 0$ and $n\geq 1$. Now to show that $exp_p(p^k(A/H-\#|A/H|))$ is well defined, we need

$$\nu_p\left(\frac{\{p^k(A/H - \#|A/H|)\}^n}{n!}\right) \longrightarrow \infty \text{ as } n \longrightarrow \infty.$$

The above calculation shows that this is guaranteed when

$$(p-1)(k+\nu_p(\#|A/H|)) > 1.$$

Since $\nu_p(\#|A/H|) \geq 1$, this is satisfied in particular when p is odd, or when $p=2$ and $k\geq 1$, or when $p=2$ and $\nu_p(\#|A/H|) \geq 2$. We note that under this condition

$$\nu_p\left(\frac{\{p^k(A/H - \#|A/H|)\}^n}{n!}\right) \geq k+\frac{n+\alpha(n)-2}{p-1} \qquad \cdots(\$)$$

Thus exp_p is well-defined on the basis elements of J, which is sufficient to show that exp_p is well-defined. The fact $exp_p(J) \subseteq 1+J$ follows from (\$).

Proof of (ii). Given any $x \in I(A)\hat{_p}$, we may express

$$x = \sum a_i(A/H_i - \#|A/H_i|) \text{ with } \nu_p(a_i)\rightarrow\infty \text{ as } i\rightarrow\infty.$$

Then we get

$$x^p \in \Big(\sum a_i{}^p(-\#|A/H_i|)^{p-1}(A/H_i - \#|A/H_i|)\Big) + pI(A)\hat{_p} = pI(A)\hat{_p}.$$

This implies $x^p = px_1$ for some $x_1 \in I(A)\hat{}_p$. Continuing this, we see

$$x^{p^k} = (x^{p^{k-1}})^p = \left(p^{(p^{k-2}+p^{k-3}+\ldots+1)} x_{k-1}\right)^p$$

$$= p^{(p^{k-2}+p^{k-3}+\ldots+1)p} (px_k) = p^{(p^{k-1}+p^{k-2}+\ldots+1)} x_k.$$

Now given any n, we denote $k(n) = [\mathrm{Log}_p n]$, the largest natural number such that $p^{k(n)} \leq n$. Then we have

$$\nu_p\left(\frac{x^n}{n}\right) \geq \nu_p\left(\frac{x^{p^{k(n)}}}{p^{k(n)}}\right) \geq (p^{k(n)-1} + p^{k(n)-2} + \ldots + 1) - k(n),$$

which clearly goes to ∞ as $n \to \infty$, and satisfies ≥ 0 when $k(n) \geq 1$. This implies the well-definedness of log_p.

Proof of (iii). This follows from ($).

Proof of (iv). This follows from $\frac{x^n}{n} \in S$ for any n, which follows from, just like the proof of (*ii*), the fact $x^p \in pS$ for any $x \in S$.

Proof of (v). Assuming (*i*)(*ii*)(*iv*), this is essentially well-known.

Proof of (vi). Let $x = \sum a_i(A/H_i - \#|A/H_i|)$ be the basis representation of an element $x \in S$ (so $a_i = p^{k_i} b_i$ for some $b_i \in \hat{\mathbb{Z}}_p$). We begin from the case of exp_p. We first note that

$$exp_p\Big(a_i(A/H_i - \#|A/H_i|)\Big) \in 1 + a_i(A/H_i - \#|A/H_i|) + pS,$$

which easily follows from ($) and the fact $a_i(A/H_i - \#|A/H_i|) \in S$.

Now

$$exp_p(x) = exp_p\Big(\sum a_i(A/H_i - \#|A/H_i|)\Big) = \prod exp_p\Big(a_i(A/H_i - \#|A/H_i|)\Big)$$

$$\in \prod\Big(1 + a_i(A/H_i - \#|A/H_i|) + pS\Big) \subseteq 1 + x + pS,$$

where the last inclusion follows from $x = \sum a_i(A/H_i - \#|A/H_i|)$ and the very fact that S is an exponentiable module.

We now turn to the case of log_p. Since $x \in J$,

$$\nu_p\Big(a_i{}^p(-\#|A/H_i|)^{p-1}\Big) \geq \begin{cases} 2 & \text{if } p=2 \text{ and } \#|A/H_i|\geq 4, \text{ or if } p=\text{odd} \\ 3 & \text{if } p=2 \text{ and } \#|A/H_i|=2 \end{cases},$$

(we note that, of course, this estimate is not sharp, just sufficient for the proof) which implies

$$\Big(a_i(A/H_i-\#|A/H_i|)\Big)^p \in p^2S.$$

Then, since $x\in S$,

$$x^p = \Big(\sum a_i(A/H_i-\#|A/H_i|)\Big)^p \in \sum\Big(a_i(A/H_i-\#|A/H_i|)\Big)^p + p(pS) = p^2S.$$

Arguing just like (ii), we deduce

$$\frac{x^n}{n} \in pS \text{ if } p\leq n.$$

Since it is trivial to see

$$\frac{x^n}{n} \in pS \text{ for } 2\leq n\leq p-1,$$

this completes the proof. □

Corollary 2.10. (i) *For any* $x \in I(A)\hat{}_p$, $(1+x)^p \in 1+pI(A)\hat{}_p$. *Furthermore if* x *is contained in an exponentiable module* S, *then* $(1+x)^p \in 1+pS$.
(ii) *For any* $y \in J$, $1+py \in (1+J)^p$. *Furthermore if* y *is contained in an exponentiable module* S, *then* $1+py \in (1+S)^p$.

Proof. (i) follows from $x^p \in pI(A)\hat{}_p$ *(resp.* $x^p\in pS$*) if* $x \in I(A)\hat{}_p$ *(resp. if* $x \in S$) (cf. the proof of Theorem 2.9.(ii)). For (ii),

$$\begin{aligned} 1+py &= exp_p\circ log_p(1+py) && \text{by 2.9.}(v) \\ &= exp_p(py+pa) && \text{for some } a\in J \text{ (or } S \text{ in the latter case) by 2.9.}(vi) \\ &= exp_p(y+a)^p \end{aligned}$$

$$=(1+b)^p \qquad \text{for some } b\in J \ (\textit{or } S \text{ in the latter case}) \text{ by } 2.9.(iii)$$ □

Theorem 2.11. *Suppose S is an exponentiable module with additive $\hat{\mathbb{Z}_p}$-basis*

$$p^{k_i}(A/H_i - \#|A/H_i|), \quad k_i\geq 0,$$

then $1+S$ has

$$\left\{1+(A/H_i - \#|A/H_i|)\right\}^{k_i}$$

as a multiplicative $\hat{\mathbb{Z}_p}$-basis.

Proof. From 2.9.(*iii*),(*iv*), $1+S$ has $exp_p\left\{p^{k_i}(A/H_i - \#|A/H_i|)\right\}$ as a multiplicative $\hat{\mathbb{Z}_p}$-basis. Now we have

$$exp_p\left\{p^{k_i}(A/H_i - \#|A/H_i|)\right\}$$

$$= exp_p(A/H_i - \#|A/H_i|)^{p^{k_i}} \qquad \text{by } 2.9.(i)$$

$$= \left\{1+(A/H_i-\#|A/H_i|)+pa(A/H_i-\#|A/H_i|)\right\}^{p^{k_i}} \qquad \text{for some } a\in\hat{\mathbb{Z}_p} \text{ by } 2.9.(vi)$$

$$=\left\{1+(A/H_i-\#|A/H_i|)\right\}^{p^{k_i}}\left\{1+pb(A/H_i-\#|A/H_i|)\right\}^{p^{k_i}} \qquad \text{for some } b\in\hat{\mathbb{Z}_p}$$

$$= \left\{1+(A/H_i- \#|A/H_i|)\right\}^{p^{k_i}}\left(\left\{1+pb(A/H_i- \#|A/H_i|)\right\}^{p^{k_i-1}}\right)^p$$

Here we see

$$\left\{1+pb(A/H_i- \#|A/H_i|)\right\}^{p^{k_i-1}}$$

$= exp_p \circ log_p\left\{1+pb(A/H_i - \#|A/H_i|)\right\}^{p^{k_i-1}}$ by 2.9.(v)

$= exp_p\left(p^{k_i-1} log_p\left\{1+pb(A/H_i - \#|A/H_i|)\right\}\right)$ by 2.9.(ii)

$= exp_p\left(p^{k_i-1}\left\{pb(A/H_i - \#|A/H_i|)+pc(A/H_i - \#|A/H_i|)\right\}\right)$

for some $c \in \hat{\mathbb{Z}_p}$ by 2.9.(vi)

$\in exp_p(S) \subseteq 1+S$ by 2.9.(iii),

which implies that

$$\left(\left\{1+pb(A/H_i - \#|A/H_i|)\right\}^{p^{k_i-1}}\right)^p \in (1+S)^p.$$

Clearly, this proves the theorem. □

Theorem 2.12. *Suppose* $x_i \in I(A)\hat{_p}$ $(i=1,2,\ldots,l)$ *satisfy* $x_i x_j \in pJ$ *for* $i \neq j$ (*resp.* x_i *are contained in an exponentiable module* S), *then*

$$1 + \sum_{i=1}^{l} x_i = \left(\prod_{i=1}^{l}(1+x_i)\right)\times(1+y)^p \quad \textit{for some} \quad y \in J \ (\textit{resp. for some } y \in S).$$

Proof. This is a straight forward consequence of Corollary 2.10.(ii):

$\prod_{i=1}^{l}(1+x_i)$

$= 1 + \sum_{i=1}^{l} x_i + pa$ for some $a \in J$ (*resp.* $a \in S$) from the assumption.

$= \left(1 + \sum_{i=1}^{l} x_i\right)\left\{1 + pa\left(1 + \sum_{i=1}^{l} x_i\right)^{-1}\right\}$

$$= \Big(1 + \sum_{i=1}^{l} x_i\Big)(1 + pb) \qquad b = a\Big(1 + \sum_{i=1}^{l} x_i\Big)^{-1} \in J.$$

$(resp.\ \in S)$.

$$= \Big(1 + \sum_{i=1}^{l} x_i\Big)(1+c)^p$$

for some $c \in J$ $(resp. \in S)$ from Corollary 2.10.

Therefore, taking $y \in J$ $(resp.\ y \in S)$ so that $1+y = (1+c)^{-1}$, we get

$$1 + \sum_{i=1}^{l} x_i = \Big(\prod_{i=1}^{l}(1+x_i)\Big)\times(1+y)^p$$

as was desired. □

Corollary 2.13. *Suppose $x \in J$ (resp. $x \in S$: exponentiable module) satisfies $x^2 \in pJ$ resp. $x^2 \in pS$), then*

$$(1+x)^{-1} = (1+x)^{p-1}(1+z)^p \quad \text{for some } z \in J\ (resp.\ z \in S).$$

Proof. This is a straight-forward consequence of Corollary 2.10. (*ii*), and Theorem 2.12., since

$$\begin{aligned}
(1+x)^{-1} &= 1-x+x^2-x^3+x^4-x^5+\cdots \\
&= 1-x+x^2(1+x)^{-1} \\
&= (1-x)\Big\{1+x^2(1-x^2)^{-1}\Big\} \\
&= \Big\{1+(p-1)x-px\Big\}\Big\{1+x^2(1-x^2)^{-1}\Big\} \\
&= \Big\{1+(p-1)x\Big\}\Big\{1-px\{1+(p-1)x\}^{-1}\Big\}\Big\{1+x^2(1-x^2)^{-1}\Big\},
\end{aligned}$$

where $-px\{1+(p-1)x\}^{-1}$, $x^2(1-x^2)^{-1} \in pJ$ $(resp. \in pS)$. □

We now begin to write down the elements in question multiplicatively. Of course, this is to get important relations in $H_*(SG)$ and its multiplicative homology operations, and the following Prop. 2.14 (*resp.* Prop. 2.15) will play the key role in the proof of Th. 3.14 (*resp.* Th. 3.19). In the both propositions, the first 2 terms will be impotant and the latter 2 terms's contribution in homology will be absorbed in some appropriate modulus.

Proposition 2.14. *Suppose* $r \geq 2$, *then*

$$\mathrm{Res}^{\Sigma_p \int (\mathbb{Z}/p)^r}_{(\mathbb{Z}/p)\times(\mathbb{Z}/p)^r} \mathcal{P}^p_{\otimes}((\mathbb{Z}/p)^r + (1-p^r))$$

$$= \left\{1 + \left\{(\mathbb{Z}/p)\times(\mathbb{Z}/p)^r - p^{r+1}\right\}\right\}$$

$$\times \left\{ \prod_{\substack{|H_i| = p \\ H_i \subset (\mathbb{Z}/p)\times(\mathbb{Z}/p)^r}} \left\{1 + \left(((\mathbb{Z}/p)\times(\mathbb{Z}/p)^r)/H_i - p^r\right)\right\}^{a_i} \right\}$$

$$\times \left\{1 + \left\{p\, ((\mathbb{Z}/p)\times(\mathbb{Z}/p)^r)/(\mathbb{Z}/p)^r - p)\right\}\right\}^C \times \{1+s\}^p,$$

for some $c \in \mathbb{Z}$ *and* $s \in S$, *where* S *is an exponentiable module generated by*

$$\left\{(\mathbb{Z}/p)\times(\mathbb{Z}/p)^r - p^{r+1}\right\},\ \left\{((\mathbb{Z}/p)\times(\mathbb{Z}/p)^r)/H_i - p^r\right\}_{|H_i| = p},$$

and

$$\left\{p\, ((\mathbb{Z}/p)\times(\mathbb{Z}/p)^r)/(\mathbb{Z}/p)^r - p)\right\}.$$

Proof. From Proposition 2.1,

$$\mathrm{Res}^{\Sigma_p \int (\mathbb{Z}/p)^r}_{(\mathbb{Z}/p)\times(\mathbb{Z}/p)^r} \mathcal{P}^p_{\otimes}((\mathbb{Z}/p)^r + (1-p^r))$$

$$= \left\{\mathrm{Res}^{\Sigma_p\int(\mathbb{Z}/p)^r}_{(\mathbb{Z}/p)\times(\mathbb{Z}/p)^r}\mathfrak{P}^p_\otimes((\mathbb{Z}/p)^r)\right\} + \left\{\mathrm{Res}^{\Sigma_p\int(\mathbb{Z}/p)^r}_{(\mathbb{Z}/p)\times(\mathbb{Z}/p)^r}\mathfrak{P}^p_\otimes(1-p^r)\right\}$$

$$+ \mathrm{Res}^{\Sigma_p\int(\mathbb{Z}/p)^r}_{(\mathbb{Z}/p)^{r+1}}\left(\sum_{i=1}^{p-1}\mathrm{Ind}^{\oplus}{}^{\Sigma_p\int(\mathbb{Z}/p)^r}_{\Sigma_i\int(\mathbb{Z}/p)^r\times\Sigma_{p-i}\int(\mathbb{Z}/p)^r}(\mathfrak{P}^i_\otimes((\mathbb{Z}/p)^r)\times\mathfrak{P}^{p-i}_\otimes(1-p^r))\right)$$

The first term was calculated in Lemma 2.5, the second in Lemma 2.6, amd the third in Corollary 2.4. So, this is equal to

$$= \left\{l\,((\mathbb{Z}/p)\times(\mathbb{Z}/p)^r) + \sum_{\substack{|H_i|=p\\ H_i\subset(\mathbb{Z}/p)\times(\mathbb{Z}/p)^r}} a_i\,((\mathbb{Z}/p)\times(\mathbb{Z}/p)^r)/H_i\right\}$$

$$+ \left\{(1-p^r) + m\,((\mathbb{Z}/p)\times(\mathbb{Z}/p)^r)/(\mathbb{Z}/p)^r\right\} + \left\{k\,((\mathbb{Z}/p)\times(\mathbb{Z}/p)^r)\right\},$$

where

$$k = \frac{1-(p^r)^p-(1-p^r)^p}{p\times p^r} \equiv 1 \ (\mathrm{mod.}\ p),\quad l = p^{r-1}((p^r)^{p-2}-1) \equiv 0 \ (\mathrm{mod.}\ p),$$

$$m = \frac{(1-p^r)^p-(1-p^r)}{p} \equiv 0 \ (\mathrm{mod.}\ p) \ (\text{This follows from } r\geq 2),$$

and $a_i\in\mathbb{Z}$ is some integer given in Corollary 2.4.

This implies that

$$\mathrm{Res}^{\Sigma_p\int(\mathbb{Z}/p)^r}_{(\mathbb{Z}/p)\times(\mathbb{Z}/p)^r}\mathfrak{P}^p_\otimes((\mathbb{Z}/p)^r+(1-p^r))$$

$$= 1 + \left\{(\mathbb{Z}/p)\times(\mathbb{Z}/p)^r - p^{r+1}\right\}$$

$$+ \sum_{\substack{|H_i|=p\\ H_i\subset(\mathbb{Z}/p)\times(\mathbb{Z}/p)^r}} a_i\left\{((\mathbb{Z}/p)\times(\mathbb{Z}/p)^r)/H_i - p^r\right\}$$

$$+ c\left\{p\,((\mathbb{Z}/p)\times(\mathbb{Z}/p)^r)/(\mathbb{Z}/p)^r - p)\right\} + p\left\{d((\mathbb{Z}/p)\times(\mathbb{Z}/p)^r - p^{r+1})\right\}$$

for some a_i, c, $d \in \mathbb{Z}$.

Now the point is that

$$\left\{(\mathbb{Z}/p)\times(\mathbb{Z}/p)^r - p^{r+1}\right\},\ \left\{((\mathbb{Z}/p)\times(\mathbb{Z}/p)^r)/H_i - p^r\right\}_{|H_i|=p},$$

and

$$\left\{p\ ((\mathbb{Z}/p)\times(\mathbb{Z}/p)^r)/(\mathbb{Z}/p)^r - p)\right\}$$

form an exponentiable module S. This follows from Example 2.8.(*ii*)(*iii*). Therefore we can apply Theorem 2.12 to get

$$\mathrm{Res}^{\Sigma_p\int(\mathbb{Z}/p)^r}_{(\mathbb{Z}/p)\times(\mathbb{Z}/p)^r}\mathcal{P}^p_{\otimes}((\mathbb{Z}/p)^r+(1-p^r))$$

$$= \left\{1 + \left\{(\mathbb{Z}/p)\times(\mathbb{Z}/p)^r - p^{r+1}\right\}\right\}$$

$$\times\left\{\prod_{\substack{|H_i|=p \\ H_i\subset(\mathbb{Z}/p)\times(\mathbb{Z}/p)^r}}\left\{1 + \left(((\mathbb{Z}/p)\times(\mathbb{Z}/p)^r)/H_i - p^r\right)\right\}^{a_i}\right\}$$

$$\times\left\{1 + \left\{p\ ((\mathbb{Z}/p)\times(\mathbb{Z}/p)^r)/(\mathbb{Z}/p)^r - p)\right\}\right\}^C \times \{1+s\}^P,$$

for some $s\in S$, as was desired. □

Proposition 2.15. (*i*) *Suppose p = odd prime. Put $G=(\mathbb{Z}/p)^a$ and $H_{j_1j_2,..,j_k} = (\mathbb{Z}/p)\times*\cdots\times*\times\cdots*\times(\mathbb{Z}/p) \subset G$, where $*$ appears in the $j_1,j_2,..,j_k$ -th coordinates $(1\le j_1<j_2,..,<j_k\le a)$ and $2\le a\le p$. (Therefore, as a group, $H_{j_1j_2,..,j_k} \cong (\mathbb{Z}/p)^{a-k}$.) Then we get*

$$
\begin{aligned}
&1+\sum_{1\le j\le a}(G/H_j-p)\\
&=\left\{\prod_{1\le j\le a}\left\{1+(G/H_j-p)\right\}\right\}\\
&\times\prod_{k=2}^{a-1}\left(\prod_{1\le j_1<\dots<j_k\le a}\left\{1+(G/H_{j_1j_2,\dots,j_k}-p^k)\right\}\right)^{\alpha(k)}\\
&\times\left\{1+(G-p^a)\right\}\times(1+v)^p
\end{aligned}
$$

for some $v\in J=I(G)\hat{}_p$, where $\alpha(k)$ is chosen so that $1\le\alpha(k)\le p-1$ and $\alpha(k)\equiv(-1)^{k+1}(k-1)!$ (mod.p).

(ii) Suppose $p=2$. *Put* $G=(\mathbb{Z}/2)\times(\mathbb{Z}/2)$, $H_1=*\times(\mathbb{Z}/2)\subset G$, *and* $H_2=(\mathbb{Z}/2)\times*\subset G$. *Then we get*

$$
\begin{aligned}
1+(G/H_1-2)+(G/H_1-2)&=\left\{\left\{1+(G/H_1-2)\right\}\left\{1+(G/H_2-2)\right\}\right\}\\
&\times\left\{1+2(G/H_1-2)\right\}\times\left\{1+2(G/H_2-2)\right\}\\
&\times\left\{1+(G-4)\right\}\times(1+l)^2
\end{aligned}
$$

for some $l\in J=D(G)\hat{}_2+2I(G)\hat{}_2$.

Proof. (i) For simplicity, we put $t_j=(G/H_j-p)$. Then

$$
\begin{aligned}
&1+\sum_{1\le j\le a}(G/H_j-p)\\
&=1+\sum_{j=1}^{a}t_j\\
&=exp_p\circ log_p\Big(1+\sum_{j=1}^{a}t_j\Big) &&\text{by Theorem 2.9.}(i)\\
&=exp_p\left\{\sum_{k\ge1}(-1)^{k+1}\frac{\Big(\sum_{j=1}^{a}t_j\Big)^k}{k}\right\}.
\end{aligned}
$$

Now from the proof of Theorem 2.9.(i),

$$\nu_p\left\{\frac{\left(\sum_{j=1}^{a} t_j\right)^k}{k}\right\} \geq p-1 \qquad \text{if } k\geq p^2. \tag{1}$$

On the other hand, since $a\leq p$,

$$\frac{\left(\sum_{j=1}^{a} t_j\right)^p}{p} = \begin{cases} \dfrac{\left(\sum_{j=1}^{a} t_j{}^p\right)}{p} + \sum_j a_j(t_j)^2 & \text{if } 2\leq a\leq p-1 \\ \dfrac{\left(\sum_{j=1}^{a} t_j{}^p\right)}{p} + \sum_j b_j(t_j)^2 + (p-1)!\,(t_j t_2\cdots t_p) & \text{if } a=p \end{cases}$$

for some a_j, $b_j\in I(G)\hat{}_p$. As

$$t_j{}^2 = (G/H_j-p)^2 = p(G/H_j-p)-2p(G/H_j-p) = -p(G/H_j-p) = -pt_j\,,$$

this implies that

$$\nu_p\left\{\frac{\left(\sum_{j=1}^{a} t_j\right)^k}{k}\right\} \geq 1 \qquad \text{if } a+1\leq k<p^2. \tag{2}$$

From (1) and (2), we can write

$$\sum_{k\geq 1} (-1)^{k+1}\frac{\left(\sum_{j=1}^{a} t_j\right)^k}{k} = \sum_{k=1}^{a} (-1)^{k+1}\frac{\left(\sum_{j=1}^{a} t_j\right)^k}{k} + ps$$

for some $s\in I(G)\hat{}_p=J$. From this, let's consider only those k such that $k\leq a$ for the time being. Then, from $t_j{}^2 = -pt_j$, we get

$$\frac{\left(\sum_{j=1}^{a} t_j\right)^k}{k}$$

$$= (k-1)! \sum_{1\leq j_1<\cdots<j_k\leq a} t_{j_1}t_{j_2}\cdots t_{j_k} + ps'$$

for some $s'\in I(G)\hat{}_p=J$

$$= (k-1)! \sum_{1\leq j_1<\dots<j_k\leq a} (G/H_{j_1}-p)(G/H_{j_2}-p)\cdots(G/H_{j_k}-p) + ps'$$

$$= (k-1)! \sum_{1\leq j_1<\dots<j_k\leq a} \left\{(G/H_{j_1 j_2,\dots,j_k} - p^k) + ps_{j_1 j_2,\dots,j_k}\right\} + ps'$$

where s' and $s_{j_1 j_2,\dots,j_k}$ are some elements in $I(G)\hat{_p}=J$. Therefore, setting $\alpha(k)$ to be the unique natural integer such that

$$1\leq\alpha(k)\leq p-1 \quad \text{and} \quad \alpha(k)\equiv(-1)^{k+1}(k-1)! \pmod{p},$$

(note that $\alpha(1)=1$ and $\alpha(p)=p-1$) we have the following presentation:

$$\sum_{k\geq 1} (-1)^{k+1}\frac{\left(\sum_{j=1}^{a} t_j\right)^k}{k} = \sum_{k=1}^{a} \alpha(k) \left(\sum_{1\leq j_1<\dots<j_k\leq a} (G/H_{j_1 j_2,\dots,j_k} - p^k)\right) + pu$$

for some $u\in I(G)\hat{_p}=J$. Now we go back to the original position. Then, using this presentation, we have

$$1 + \sum_{1\leq j\leq a} (G/H_j-p)$$

$$= exp_p\left\{ \sum_{k\geq 1} (-1)^{k+1}\frac{\left(\sum_{j=1}^{a} t_j\right)^k}{k} \right\}$$

$$= exp_p\left\{ \sum_{k=1}^{a} \alpha(k) \left(\sum_{1\leq j_1<\dots<j_k\leq a} (G/H_{j_1 j_2,\dots,j_k} - p^k)\right) + pu\right\}$$

$$= \prod_{k=1}^{a} \left(\prod_{1\leq j_1<\dots<j_k\leq a} exp_p(G/H_{j_1 j_2,\dots,j_k} - p^k)\right)^{\alpha(k)} \times exp_p(u)^p.$$

Since each $(G/H_{j_1 j_2,\dots,j_k} - p^k)$ forms an exponentiable module, we an use Theorem 2.9.(*vi*) and Corollary 2.10.(*ii*) to express

$$exp_p(G/H_{j_1 j_2,\ldots,j_k} - p^k) = \left\{1 + (G/H_{j_1 j_2,\ldots,j_k} - p^k)\right\}(1+v_{j_1 j_2,\ldots,j_k})^p,$$

for some $v_{j_1 j_2,\ldots,j_k} \in I(G)\hat{_p} = J$. Thus

$$1 + \sum_{1 \leq j \leq a} (G/H_j - p)$$

$$= \left\{\prod_{1 \leq j \leq a} \left\{1 + (G/H_j - p)\right\}\right\}$$

$$\times \prod_{k=2}^{a-1} \left(\prod_{1 \leq j_1 < \cdots < j_k \leq a} \left\{1 + (G/H_{j_1 j_2,\ldots,j_k} - p^k)\right\} \right)^{\alpha(k)}$$

$$\times \left\{1 + (G - p^a)\right\}^{\alpha(a)} \times (1+v)^p$$

for some $v \in J = I(G)\hat{_p}$, as was desired.

Proof. (*ii*) For simplicity, we put $t_j = (G/H_j - 2)$ $(j=1,2)$ and $t_0 = (G-4)$. Then

$$\left\{\left\{1 + (G/H_1 - 2)\right\}\left\{1 + (G/H_2 - 2)\right\}\right\}$$

$$= (1+t_1)(1+t_2)$$

$$= 1 + t_1 + t_2 + t_1 t_2$$

$$= (1+t_1+t_2)\left\{1 + t_1 t_2 (1+t_1+t_2)^{-1}\right\}.$$

To examine $t_1 t_2 (1+t_1+t_2)^{-1}$, we write down the table of the number of fixed points by various subgroups of G, which completely describe $I(G)$:

	H_1	H_2	Δ	G
t_0	-4	-4	-4	-4
t_1	0	-2	-2	-2
t_2	-2	0	-2	-2
$1+t_1+t_2$	-1	-1	-3	-3
$(1+t_1+t_2)^{-1}$	-1	-1	$-\frac{1}{3}$	$-\frac{1}{3}$
$t_1t_2(1+t_1+t_2)^{-1}$	0	0	$-\frac{4}{3}$	$-\frac{4}{3}$

Here we denote by $\Delta \subset (\mathbb{Z}/2)\times(\mathbb{Z}/2)\cong G$ the diagonal subgroup which is isomorphic to $\mathbb{Z}/2$. The reader might have noticed that the trivial group $\{e\}$ doesn't appear in the list of subgroups of G, this is because our focus is on the 2-adically completed augmentation ideal $I(G)\hat{_2}$.

Now from this table, we see immediately that

$$t_1t_2(1+t_1+t_2)^{-1} = \frac{1}{3}(2t_1+2t_2-t_0)$$

$$= \frac{1}{3}\Big\{2(G/H_1-2)+2(G/H_2-2)-(G-4)\Big\}.$$

Since $I(G)\hat{_2}=I(G)\underset{\mathbb{Z}}{\otimes}\mathbb{Z}\hat{_2}$ and

$$\frac{1}{3} = \frac{1}{1+2} = 1-2+2^2-2^3+2^4-\cdots \in 3+8\mathbb{Z}\hat{_2},$$

we can deduce that

$$t_1t_2(1+t_1+t_2)^{-1} = 2(G/H_1-2)+2(G/H_2-2)+(G-4)+4r$$

for some $r\in I(G)\hat{_2}$. Again the point is the fact that the elements

$$2(G/H_1-2),\ 2(G/H_2-2),\ (G-4),\text{ and } 4I(G)\hat{_2}$$

form an exponentiable module. Then we can apply Corollary 2.11 and Corollary 2.10 (note that $4I(G)\hat{_2}\subset 2J$) to express

$$1+t_1t_2(1+t_1+t_2)^{-1}$$

$$= \Big\{1 + 2(G/H_1-2)\Big\}\Big\{1 + 2(G/H_2-2)\Big\}\Big\{1+(G-4)\Big\}(1+k)^2$$

for some $k\in I(G)\hat{_2}$.

For our purpose, we should write down the inverse of each element. This is carried out by Corollary 2.13. Therefore, summarizing our calculations, we get

$$1+(G/H_1-2)+(G/H_1-2)$$

$$= 1 +t_1+t_2$$

$$= \Big\{(1+t_1)(1+t_2)\Big\}\Big\{1+t_1t_2(1+t_1+t_2)^{-1}\Big\}^{-1}$$

$$= \Big\{\Big\{1 + (G/H_1-2)\Big\}\Big\{1 + (G/H_2-2)\Big\}\Big\}$$

$$\times\Big\{\Big\{1 + 2(G/H_1-2)\Big\}\Big\{1 + 2(G/H_2-2)\Big\}\Big\{1+(G-4)\Big\}(1+k)^2\Big\}^{-1}$$

$$= \left\{\left\{1 + (G/H_1-2)\right\}\left\{1 + (G/H_2-2)\right\}\right\}$$

$$\times \left\{1 + 2(G/H_1-2)\right\} \times \left\{1 + 2(G/H_2-2)\right\}$$

$$\times \left\{1 + (G-4)\right\} \times (1+l)^2$$

for some $l \in J$, as was desired. □

§3. HOMOLOGY OPERATIONS AND A MULTIPLICATIVE TRANSFER THEOREM

In this section all homology is taken mod p. We should write down the known results on $H_*(Q_0S^0)$ and $H_*(SG)$. For this purpose we follow the treatment of [9]. We begin by recalling the structure of $H_*(E\mathbb{Z}/p \times_{\mathbb{Z}/p} X^p)$. Let $e_i \in H_*(B\mathbb{Z}/p) \cong \mathbb{Z}/p$ be a generator and $x_0=1,x_1,x_2,\ldots..$ be basis of $H_*(X)$. Then $H_*(E\mathbb{Z}/p \times_{\mathbb{Z}/p} X^p)$ has a basis of the form

$$e_i \int x_j = e_i \otimes x_j \otimes \cdots \otimes x_j = e_i \otimes x_j{}^p \quad (i,j \geq 0)$$
$$x_{i_1}|\cdots|x_{i_p} = e_0 \otimes x_{i_1} \otimes \cdots \otimes x_{i_p} \quad (i_k \neq i_l \text{ for some } k,l)$$

where $(i_1,\ldots,i_p)$ runs through one representative of each class obtained by cyclic permutations of the indices. Remarking that $E\mathbb{Z}/p \times_{\mathbb{Z}/p} (BH)^p \simeq B(\mathbb{Z}/p \int H)$, we get the following elements in the homology of $\mathcal{G}(p^k,p)$: the k-fold iterated wreath product $\mathbb{Z}/p \int \cdots \int \mathbb{Z}/p$, which turns out to be a p-Sylow subgroup of Σ_{p^k} (see the beginning of §2 for the standard embedding); for a sequence $I=(i_1,\ldots,i_k)$ of non-negative integers we set

$$\hat{e}_I = e_{i_1} \int \cdots \int e_{i_k} \in H_*(B\mathcal{G}(p^k,p)).$$

Since we make full use of the group homology, we prepare some notations in group-homology: When G is a finite group with a finite subgroup K, we denote by

$$i_{K \to G} \colon H_*(BK) \to H_*(BG)$$

and

$$tr_{G \to K} \colon H_*(BG) \to H_*(BK)$$

the inclusion and the transfer homomorphism respectively. Furthermore we denote

$$e * f = (i_{m,n})_*(e \otimes f) \in H_*(B\Sigma_{m+n}),$$

for any $e \in H_*(B\Sigma_m)$ and $f \in H_*(B\Sigma_n)$ where $i_{m,n}$: $\Sigma_m \times \Sigma_n \to \Sigma_{m+n}$ is the usual inclusion. Then, using our fixed inclusion $\mathcal{I}(p^k,p) \to \Sigma_{p^k}$ (see the beginning of §2 for this standard embedding), we set

$$e_I = i_{\mathcal{I}(p^k,p) \to \Sigma_{p^k}}(\hat{e}_I) \in H_*(B\Sigma_{p^k}).$$

We are going to recall some results which involves the sequence I in group homology. We begin with a definition.

Definition 3.1. Consider the following three conditions for a sequence $I=(i_1,\ldots,i_k)$ and e_I:

(*i*) $i_j > 0$ $(j=1,2,\ldots,k)$,

(*ii*) $i_1 \le i_2 \le \ldots \le i_k$,

(*iii*) (*a*) $i_t \equiv 0$ or -1 mod $2(p-1)$ if the dimension of $e_{i_{t+1}} \int \ldots \int e_{i_k}$ is even, or

(*iii*) (*b*) $i_t \equiv p-1$ or $p-2$ *mod* $2(p-1)$ if the dimension of $e_{i_{t+1}} \int \ldots \int e_{i_k}$ is odd.

I is called <u>*admissible*</u> if the above three conditions are all satisfied. I is called <u>*weakly admissible*</u> if only (*ii*) and (*iii*) are satisfied. The <u>*length* $l(I)$</u> of I is defined to be k. (*So* $e_I \in H_*(B\Sigma_{p^k}) \Rightarrow l(I) \le k$)

Proposition 3.2. (*Well known and easy, see e.g.* [9, 3.7.]) *Let* $(\mathbb{Z}/p)^k \subseteq \mathcal{I}(p^k,p)$ *be the sub elementary abelian p-group of* $\mathcal{I}(p^k,p)$, *which was given at the beginning of* §2 *or* [9, *p*.95]. *Then for any* $x \in H_*(B(\mathbb{Z}/p)^k)$,

$$i_{(\mathbb{Z}/p)^k \to \mathcal{I}(p^k,p)}(x)$$

is a linear combination of elements of the form $\hat{e}_I$ *with* $l(I)=k$.

Proposition 3.3. (*Well known and easy, see e.g.* [9, 3.4.*and* 3.5.]) (*i*) *Under the same hypothesis as that of Proposition* 3.2, *for any length-k sequence I of non-negative integers, there exists some* $f_I \in H_*(B(\mathbb{Z}/p)^k)$ *such that*

$$i_{(\mathbb{Z}/p)^k \to \mathcal{G}(p^k,p)}(f_I) = \hat{e}_I \in H_*(B\mathcal{G}(p^k,p)).$$

(ii) More generally, for any $x \in H_*(X)$ *and* $i \geq 0$*, there exists some* $y \in H_*(B\mathbb{Z}/p \times X)$ *such that*

$$(1 \times \Delta)_*(y) = e_i \int x \in H_*(E\mathbb{Z}/p \underset{\mathbb{Z}/p}{\times} X^p).$$

If deg $x > 0$*, then we can take*

$$y = \sum e_j \otimes x_j \in H_*(B\mathbb{Z}/p) \otimes H_*(X) \cong H_*(B\mathbb{Z}/p \times X)$$

so that deg $x_j > 0$ *for any j. In particular, this implies*

$$(pr_1)_*(y) = 0 \in H_*(B\mathbb{Z}/p),$$

where pr_1: $B\mathbb{Z}/p \times X \to B\mathbb{Z}/p$ *is the first projection. Of course the same thing can be said about (i).*

Theorem 3.4. *(Adem relation [1]) For any sequence* $J = (j_1, \ldots, j_k)$ *of non-negative integers,*

$$i_{\mathcal{G}(p^k,p) \to \Sigma_{p^k}}(\hat{e}_J) = \sum \lambda_I e_I,$$

where $\lambda_I \in \mathbb{Z}/p$ *and* e_I *runs over weakly admissible sequences.*

Theorem 3.5. *(Kahn-Priddy [9]) Let* $x = e_{i_1} * \ldots * e_{i_u} * e_{I_1} * \ldots * e_{I_v} \in H_*(B\Sigma_n)$ *with* $i_j \geq 0$ *and* $l(I_j) \geq 2$*, then*

(i) $tr_{\Sigma_{p^k} \to \mathcal{G}(p^k,p)}(x) = \hat{e}_{i_1} | \ldots | \hat{e}_{i_u} | \hat{e}_{I_1} | \ldots | \hat{e}_{I_v} + \hat{e}_x,$

where $\hat{e}_x = \sum \hat{e}_{i_1} | \ldots | \hat{e}_{i_u} | \hat{e}_{I'_1} | \ldots | \hat{e}_{I'_v}$,

the summation being taking over certain elements of the form indicated (or permutations thereof) with $l(I'_j) = l(I_j)$*, and*

(ii) $i_{\mathcal{S}(p^k,p)\to\Sigma_{p^k}}(\hat{e}_x) = (\lambda-1)x$, *where* $\lambda=[\Sigma_{p^k} : \mathcal{S}(p^k,p)]$.

Remark.3.6. We can easily obtain

$$\lambda = [\Sigma_{p^k} : \mathcal{S}(p^k,p)] \equiv \begin{cases} 1 \text{ when } k \text{ is even} \\ -1 \text{ when } k \text{ is odd} \end{cases} \mod. p.$$

Proof of Remark 3.6. For any $n\in\mathbb{N}$, we put $\mu(n) = \frac{n}{p^{\nu_p(n)}}$: non p-divisible part of n. Then we have

$$[\Sigma_{p^k} : \mathcal{S}(p^k,p)] = \prod_{1\le n\le p^k} \mu(n) = \prod_{0\le l\le k} \prod_{1\le m\le p^l,(p,m)=1} \mu(p^{k-l}m)$$

$$= \prod_{1\le l\le k} \prod_{1\le m\le p^l,(p,m)=1} \mu(m).$$

Since $(p-1)! \equiv -1$ mod.p, we see easily

$$\prod_{1\le m\le p^l,(p,m)=1} \mu(m) \equiv (-1)^{\frac{p^l-p^{l-1}}{p-1}} = (-1)^{p^{l-1}}.$$

Therefore we get

$$[\Sigma_{p^k} : \mathcal{S}(p^k,p)] \equiv \prod_{1\le l\le k} (-1)^{p^{l-1}} = (-1)^{\sum_{1\le l\le k} p^{l-1}} \mod. p.$$

Now the claim follows from the trivial fact that

$$\sum_{1\le l\le k} p^{l-1} \text{ is } \begin{Bmatrix} \text{odd when } k \text{ is odd} \\ \text{even when } k \text{ is even} \end{Bmatrix}, \text{ if } p \text{ is odd.}$$ □

Next we turn to $H_*(QS^0)$. As usual, $QS^0=\coprod_n Q_nS^0$ and $Q_1S^0=SG$. First of all we have two types (additive and multiplicative) of homology operations on $H_*(QS^0)$ as follows; for any $x\in H_*(QS^0)$,

$$Q_i(x) = \theta_*(e_i\textstyle\int x) \in H_*(QS^0)$$

$$\tilde{Q}_i(x) = \xi_*(e_i \int x) \in H_*(QS^0).$$

For convenience we denote

$$Q_I = Q_{i_1} Q_{i_2} \cdots Q_{i_k}, \text{ and } \tilde{Q}_I = \tilde{Q}_{i_1} \tilde{Q}_{i_2} \cdots \tilde{Q}_{i_k},$$

for any sequence $I=(i_1,\ldots,i_k)$ of non-negative integers. We also denote two (additive and multiplicative) Pontryagin products on $H_*(QS^0)$ as follows; for any $x,y \in H_*(QS^0)$,

$$x*y = \theta_*(x \otimes y) \in H_*(QS^0)$$

$$xy \text{ (or } x\#y) = \xi_*(x \otimes y) \in H_*(QS^0).$$

The symbol $*$ was also used in the homology of the symmetric group; in fact the two $*$'s are consistent through the Barratt-Priddy-Quillen map. So we can confuse these freely. Now we recall the Pontryagin ring structure of the related spaces.

Theorem 3.7. (*Kudo-Araki* [2] *for* $p=2$, *Dyer-Lashof* [7] *for* $p=odd$)

$$H_*(Q_0S^0) = A[Q_I[1]*[-p^{l(I)}] \mid I \text{ admissible}],$$

where $A[?]$ *denotes the graded free commutative algebra functor.*

Theorem 3.8. (*i*) (*Milgram* [16]) *For* $p=2$,

$$H_*(SG) \cong E[x_1,x_2,\ldots] \otimes \mathbb{Z}/2[x_{(0,a)}, x_I \mid a>0, \text{ or } I \text{ admissible}],$$

where $x_i = Q_i[1]*[-1]$ *and* $x_I = Q_I[1]*[1-2^{l(I)}]$ *when* $l(I) \geq 2$.

(*ii*) (*Tsuchiya* [25], *May* [5]) *For* $p=odd$,

$$H_*(SG) \cong A[x_I \mid I \text{ admissible}],$$

where $x_I = Q_I[1]*[1-p^{l(I)}]$.

Theorem 3.8 will be further refined in Theorem 3.18. Next, we recall an obvious relationship between $e_I \in H_*(B\Sigma_{p^k})$, $Q_I[1] \in H_*(Q_{p^k}S^0)$, and $x_I \in H_*(SG)$, where

$k = l(I)$:

Lemma 3.9. $$\alpha_{\Sigma_{p^k}}(\Sigma_{p^k}/\Sigma_{p^k-1})_*(e_I) = Q_I[1] \in H_*(Q_{p^k}S^0),$$

and $$\alpha_{\Sigma_{p^k}}\Big(1 + (\Sigma_{p^k}/\Sigma_{p^k-1} - p^k)\Big)_*(e_I) = x_I \in H_*(SG).$$

Proof. We only have to prove the first statement: This follows easily from the fact that e_I comes from $\hat{e}_I \in H_*(B\mathcal{I}(p^k,p))$, the additive case of Lemma1.2., and the definition of Q_I. □

Using the structure of $H_*(SG)$, we define B to be the subalgebra of $H_*(SG)$ generated by x_I with $l(I) \geq 2$ such that I admissible and when p=2 also sequences of the form $I=(0,a)$ $a \in \mathbb{N}$. We denote I_1 to be the set of positive degree elements of $H_*(SG)$ (subscript 1 of I_1 corresponds to the fact that SG is the 1-st component Q_1S^0 of QS^0), so I_1 is the augmentation ideal $\coprod_{*>0} H_*(SG)$. Then we put $\bar{B}$ to be the augmentation ideal of B, i.e. $\bar{B} = B \cap I_1$. Finally let $\mathfrak{P}$ and $\tilde{\mathfrak{P}}$ denote the p-th power operations on $H_*(QS^0)$ in the $*$ and $\#$ products respectively. We remark that

$$\tilde{\mathfrak{P}}(I_1) = \tilde{\mathfrak{P}}(\bar{B}) \subset \bar{B} \# \bar{B},$$

when $p=2$.

We now note a simple relationship between $H_*(SG)$ and group homology through

$$\alpha_G\colon A(G) \longrightarrow [BG,\ QS^0] = \pi_s^0(BG_+),$$

when $G = (\mathbb{Z}/p)^r$.

Lemma 3.10. *Let $H \subseteq G = (\mathbb{Z}/p)^r$ be a subgroup of order p^s $(s \leq r)$, then any element of*

$$\alpha_G(1 + (G/H - p^{r-s}))_*(\coprod_{*>0} H_*(BG))$$

can be written as

$$\sum \mu_I x_I + \widetilde{\mathfrak{P}}(y) \in I_1,$$

where $y \in I_1$, $\mu_I \in \mathbb{Z}/p$, *and* x_I *runs over admissible sequence with* $l(I)=r-s$ *and furthermore over* $x_{(0,a)}$ *when* $p=2$, $r-s=2$.

Proof. First of all, we note that $\alpha_G(1+(G/H-p^{r-s}))$ is homotopic to the composite

$$BG \xrightarrow{Bq} B(G/H) \xrightarrow{\alpha_{G/H}(1+(G/H-p^{r-s}))} SG,$$

where q: $G \to G/H$ is the quotient homomorphism. Now the second map factors through as follows:

$$B(G/H) \to B\mathfrak{I}(p^{r-s},p) \to B\Sigma_{p^{r-s}} \xrightarrow{\alpha_{\Sigma_{p^{r-s}}}(1+(\Sigma_{p^{r-s}}/\Sigma_{p^{r-s}-1}-p^{r-s}))} SG.$$

Therefore from Proposition 3.2., Theorem 3.4., and Lemma 3.9., any element of

$$\alpha_G(1+(G/H-p^{r-s}))_*(\coprod_{*>0} H_*(BG))$$

is a linear combination of x_I with I weak admissible. Now the difference between weak admissible and admissible lies in the sequences of the form $I=(0,J)$. The point is that, unless $p=2$ and $I=(0,a)$ $a>0$,

$$x_I = x_{(0,J)} = Q_0 Q_J[1]*[1-p^{l(I)}] = [1]*\mathfrak{P}(Q_J[1]*[-p^{l(J)}])$$

can be written as $\widetilde{\mathfrak{P}}(y)$ for some $y \in I_1$. This follows from Proposition 3.12 below. Thus the proof is completed. □

Now we are in a position to be able to apply the results, which we obtained in sections 1 and 2, to $H_*(QS^0)$. The following easy lemma, whose proof is omitted, is very fundamental in our applications.

Lemma 3.11. (*i*) *Let X be a space and Y be a homotopy associative H-space. Suppose f_i: $X \to Y$ be a continuous map ($1 \le i \le q$) and put F: $X \to Y$ to be the following composition;*

$$F: X \xrightarrow{diagonal} X^q \xrightarrow{f_1 \times f_2 \times \cdots \times f_q} Y^q \xrightarrow{m} Y,$$

where m is the product of Y. Then for any $x \in H_*(X)$,

$$F_*(x) \equiv \sum_{1 \le i \le q} (f_i)_*(x) \text{ mod. } \bar{S},$$

where S is the subring of $H_*(Y)$ generated by $J_i J_j$ with $i \ne j$ for $J_i = \mathrm{Im}((f_i)_*$: $H_*(X) \to H_*(Y)\ * > 0$). and $\bar{S}$ is its augmentation, i.e. $\bar{S} = S \cap (H_*(Y)\ * > 0)$.

(*ii*) *In addition to the setting in (i), we further suppose that Y is homotopy commutative. If we are working in the mod-p homology, $q = p^n$, and $f_1 = f_2 = \cdots = f_{p^n}$ (we denote these by f), then for any $x \in H_*(X)$, we have for some $t \in H_{\frac{*}{p^n}}(X)$,*

$$F_*(x) = \mathfrak{P}^n(f_{\frac{*}{p^n}}(t)),$$

where $\mathfrak{P}$: $H_{\frac{*}{p}}(Y) \to H_*(Y)$ is the p-th power Frobenius operation.
(Of course, when $\frac{*}{p^n} \notin \mathbb{N}$, $\mathfrak{P}^n$ isn't defined and we get $F_*(x) = 0$.)

Before we begin the applications of our theory, we must state the following result, which was used in the proof of Lemma 3.10.

Proposition 3.12. (*Madsen* [12] *for* $p=2$. *Tsuchiya* [26] *for* $p=odd$. *See also* [5]) *Let $\mathfrak{P}$ and $\tilde{\mathfrak{P}}$ denote the p-th power operations on $H_*(QS^0)$ in the $*$ and $\#$ products respectively. Then we have*

(*i*) Im $\tilde{\mathfrak{P}} \subseteq$ (Im $\mathfrak{P}$)$*[1]$.

(*ii*) $\left(\mathrm{Im}\ \mathfrak{P}|_{R*[-1]}\right)*[1] \in \mathrm{Im}\ \tilde{\mathfrak{P}}|_R = \mathrm{Im}\ \tilde{\mathfrak{P}}$,

where

$$R=\begin{cases}\mathbb{Z}/2[x_{(0,a)},\ x_I\,|\,a>0, I\ \textit{admissible}] & p=2\\ H_*(SG) & p=odd\end{cases}$$

Proof of the case $p=2$. (*ii*) We first remark that for any $y\in H_*(SG)$ there exist some elementary abelian 2-group $(\mathbb{Z}/2)^k$, $a\in H_*(B(\mathbb{Z}/2)^k)$, and $t\in I((\mathbb{Z}/2)^k)$ such that

$$\alpha_{(\mathbb{Z}/2)^k}(t)_*(a) = y*[-1] \in H_*(Q_0S^0).$$

We note that if $y\in\mathbb{Z}/2[x_{(0,a)},\ x_I\,|\,a>0,\ I$ admissible], then we can take the above t to be contained in $D((\mathbb{Z}/2)^k)^{\hat{}}_2 \subset J$ (see Definition 2.7 for these modules) with $k\geq 2$. We then claim that we can find some element $b\in H_*(B(\mathbb{Z}/2)^k)$ such that

$$(B\Delta)_*(b) = a\otimes a + \sum_\lambda (x_\lambda\otimes y_\lambda + y_\lambda\otimes x_\lambda)$$
$$\in H_*(B(\mathbb{Z}/2)^k)\otimes H_*(B(\mathbb{Z}/2)^k) = H_*(B((\mathbb{Z}/2)^k\times(\mathbb{Z}/2)^k)),$$

where Δ: $(\mathbb{Z}/2)^k \to (\mathbb{Z}/2)^k\times(\mathbb{Z}/2)^k$ is the diagonal. This is because the mod 2 cohomology ring $H^*(B(\mathbb{Z}/2)^k)$ is the polynomial ring, so the square map is injective (this is where our proof breaks down for odd prime, see Remark 3.13 below). Thus the following composition sends b to $\mathfrak{P}(y)$.

$$H_*(B(\mathbb{Z}/2)^k) \xrightarrow{(B\Delta)_*} H_*(B(\mathbb{Z}/2)^k\times B(\mathbb{Z}/2)^k)$$
$$\xrightarrow{\left(\alpha_{(\mathbb{Z}/2)^k}(t)\times\alpha_{(\mathbb{Z}/2)^k}(t)\right)_*} H_*(Q_0S^0\times Q_0S^0) \xrightarrow{\theta_*} H_*(Q_0S^0)$$

Using the additive case of Lemma 1.1, we immediately get

$$\alpha_{(\mathbb{Z}/2)^k}(2t)_*(b) = \mathfrak{P}(y),$$

which trivially implies

$$\alpha_{(\mathbb{Z}/2)^k}(1+2t)_*(b) = \mathfrak{P}(y)*[1].$$

Therefore the typical element of $\left(\mathrm{Im}\ \mathfrak{P}|_{R*[-1]}\right)*[1]$ is of the form

$$\alpha_{(\mathbb{Z}/2)^k}(1+2t)_*(b),$$

with some $k \geq 2$, $t \in D((\mathbb{Z}/2)^k)\hat{}_2$, $b \in H_*(B(\mathbb{Z}/2)^k)$. Recall that our philosophy is to work at the level of the Burnside ring until the very last moment (see the Abstract or Introduction of this paper!), so we have to study $1+2t$ now. In this case, we can apply Corollary 2.10.(*ii*) to rewrite as

$$1+2t = (1+u)^2$$

for some $u \in J = D((\mathbb{Z}/2)^k)\hat{}_2 + 2\ I((\mathbb{Z}/2)^k)\hat{}_2$. Now, thanks to the multiplicative case of Lemma 1.1, we can apply Lemma 3.11 (*ii*) to get

$$\alpha_{(\mathbb{Z}/2)^k}\left((1+u)^2\right)_*(b) = \tilde{\mathfrak{P}}\left(\alpha_{(\mathbb{Z}/2)^k}(1+u)_*(c)\right)$$

for some $c \in H_*(B(\mathbb{Z}/2)^k)$. This completes the proof.

Proof (*i*) The proof is exactly like that of (*ii*). Here we use Corollary 2.10 (*i*), the additive case of Lemma 1.1 instead of Corollary 2.10 (*ii*), the multiplicative case of Lemma 1.1 respectively. □

Remark 3.13. The original proofs of Proposition 3.12, due to Madsen and Tsuchiya (May), made use of the dual Dyer-Lashof algebra, and are thus very complicated. I hope the reader has enjoyed the above-presented very elementary and conceptual proof for the mod-2 case. But, unfortunately, we've been unable to give the proof for the odd primary version based upon our philosophy just like our proof of the mod-2 case. Of course our proof suggests that the difficulty comes from the fact that $H^*(B(\mathbb{Z}/p)^k)$ isn't a polynomial ring when p=odd, but the essential difficulty originates from the structure of $H^*(SG)$. $H^*(SG)$ isn't a free commutative algebra anymore unlike the $p=2$ case, and contains $\big(\mathbb{Z}/p[x]\big)/(x^p)$ ($\deg x$ = even) in its tensor product decomposition, which is given by applying the Borel structure theorem. This phenomenon was first observed by Nakaoka [19] in his study of $H^*(B\Sigma_\infty)$.

Now we begin our applications.

Theorem 3.14. *For any sequece I of non-negative integers with $l(I)\geq 2$,*

$$\tilde{Q}_r(x_I) \equiv x_{(r,I)} + \sum a_J x_J$$

where the congruence is taken with respect to

$$\text{mod.}\begin{cases} \bar{B}\#\bar{B} & \text{when } p=2 \\ \widetilde{\mathfrak{P}}(I_1)+\widetilde{\mathfrak{P}}(I_1)\#\widetilde{\mathfrak{P}}(I_1)+\bar{B}\#\widetilde{\mathfrak{P}}(I_1)+\bar{B}\#\bar{B} & \text{when } p=\text{odd}, \end{cases}$$

J runs over the admissible sequences with $l(J)=l(I)$ and furthermore over $(0,a)$ $a\in\mathbb{N}$ when $p=2$ and $l(I)=2$, and a_J is some element in $\mathbb{F}_p$.

Proof. We may suppose that $\deg x_I > 0$. Then from Proposition 3.3 there exists some $x\in H_*(B(\mathbb{Z}/p)^{1+l(I)})$ such that

$$i_{(\mathbb{Z}/p)^{1+l(I)}\to\mathfrak{I}(p^{1+l(I)},p)}(x) = \hat{e}_{(r,\,I)} \in H_*(B\mathfrak{I}(p^{1+l(I)},p)),$$

$$(pr_1)_*(x) = 0 \in H_*(B(\mathbb{Z}/p)),$$

where pr_1: $B(\mathbb{Z}/p)^{1+l(I)}\to B(\mathbb{Z}/p)$ is the first projection. Then, thanks to the multiplicative case of Lemma 1.2, Lemma 3.9 implies

$$\tilde{Q}_r(x_I)$$

$$= \alpha_{\Sigma_p\int\Sigma_{p^{l(I)}}}\Big(\mathfrak{P}^p_{\otimes}\big(1+(\Sigma_{p^{l(I)}}/\Sigma_{p^{l(I)-1}} - p^{l(I)})\big)\Big)_*(e_r\textstyle\int e_I)$$

$$= \alpha_{(\mathbb{Z}/p)\times(\mathbb{Z}/p)^{l(I)}}\left(\operatorname{Res}^{\Sigma_p\int(\mathbb{Z}/p)^{l(I)}}_{(\mathbb{Z}/p)\times(\mathbb{Z}/p)^{l(I)}}\mathfrak{P}^p_{\otimes}((\mathbb{Z}/p)^{l(I)}+(1-p^{l(I)}))\right)_*(x)\ .$$

Our philosophy was to study the Burnside ring element first, and in this case it was shown in Theorem 2.14 that

$$\operatorname{Res}^{\Sigma_p\int(\mathbb{Z}/p)^{l(I)}}_{(\mathbb{Z}/p)\times(\mathbb{Z}/p)^{l(I)}}\mathfrak{P}^p_{\otimes}((\mathbb{Z}/p)^{l(I)}+(1-p^{l(I)}))$$

$$= \left\{1 + \left\{(\mathbb{Z}/p)\times(\mathbb{Z}/p)^{l(I)} - p^{l(I)+1}\right\}\right\}$$

$$\times \left\{ \prod_{\substack{|H_i|=p \\ H_i\subset(\mathbb{Z}/p)\times(\mathbb{Z}/p)^{l(I)}}} \left\{1 + \left(((\mathbb{Z}/p)\times(\mathbb{Z}/p)^{l(I)})/H_i - p^{l(I)}\right)\right\}^{a_i}\right\}$$

$$\times \left\{1 + \left\{p\ ((\mathbb{Z}/p)\times(\mathbb{Z}/p)^{l(I)})/(\mathbb{Z}/p)^{l(I)} - p)\right\}\right\}^C \times \{1+s\}^p,$$

for some $c\in\mathbb{Z}$ and $s\in S$, where S is an exponentiable module generated by

$$\left\{(\mathbb{Z}/p)\times(\mathbb{Z}/p)^{l(I)} - p^{l(I)+1}\right\},\ \left\{((\mathbb{Z}/p)\times(\mathbb{Z}/p)^{l(I)})/H_i - p^{l(I)}\right\}_{|H_i|=p}\ ,$$

and

$$\left\{p\ ((\mathbb{Z}/p)\times(\mathbb{Z}/p)^{l(I)})/(\mathbb{Z}/p)^{l(I)} - p)\right\}.$$

Now, thanks to the multiplicative case of Lemma 1.1, we are in a position to apply Lemma 3.11, and the necessarilly recepies are:

$$\alpha_{(\mathbb{Z}/p)\times(\mathbb{Z}/p)^{l(I)}}\left(1 + \left\{(\mathbb{Z}/p)\times(\mathbb{Z}/p)^{l(I)} - p^{l(I)+1}\right\}\right)_*(x) = x_{(r,I)}\ ,$$

$$\alpha_{(\mathbb{Z}/p)\times(\mathbb{Z}/p)^{l(I)}}\left(1 + \left(((\mathbb{Z}/p)\times(\mathbb{Z}/p)^{l(I)})/H_i - p^{l(I)}\right)\right)_*(x) = \sum b^i{}_J x_J,$$

where J *runs over the admissible sequences with* $l(J)=l(I)$ *and furthermore over* $(0,a)$ $a\in\mathbb{N}$ *when* $p=2$ *and* $l(I)=2$, *and* $b^i{}_J$ *is some element in* $\mathbb{F}_p$,

$$\alpha_{(\mathbb{Z}/p)\times(\mathbb{Z}/p)^{l(I)}}\left(1 + \left\{p\ ((\mathbb{Z}/p)\times(\mathbb{Z}/p)^{l(I)})/(\mathbb{Z}/p)^{l(I)} - p)\right\}\right)_*(x) = 0\ ,$$

$$\alpha_{(\mathbb{Z}/p)\times(\mathbb{Z}/p)^{l(I)}}\left(\{1+s\}^p\right)_*(x) \in \widetilde{\mathfrak{P}}(I_1).$$

Here the first equality follows from the choice of x and Lemma 3.9, the second from Lemma 3.10, the third from $(pr_1)_*(x) = 0$, and the forth from Lemmma 3.11(ii) and the multiplicative case of Lemma 1.1. Of course, we are going to put $a_J = \sum_i b^i{}_J$ (we don't care about it's actual value), but our proof is still imcomplete until we deal with the modulus in the formula of Lemma 3.11(ii); the necessarilly recepie for this is:

For $*>0$,

$$\mathbf{Im}\ \alpha_{(\mathbb{Z}/p)\times(\mathbb{Z}/p)^{l(I)}}\left(1 + \left\{(\mathbb{Z}/p)\times(\mathbb{Z}/p)^{l(I)} - p^{l(I)+1}\right\}\right)_* \subseteq \bar{B}\#\widetilde{\mathfrak{P}}(I_1)\ ,$$

$$\mathrm{Im}\ \alpha_{(\mathbb{Z}/p)\times(\mathbb{Z}/p)^{l(I)}}\left(1 + \left(((\mathbb{Z}/p)\times(\mathbb{Z}/p)^{l(I)})/H_i - p^{l(I)}\right)\right)_* \subseteq \bar{B}\#\widetilde{\mathfrak{P}}(I_1)\ ,$$

$$\mathrm{Im}\ \alpha_{(\mathbb{Z}/p)\times(\mathbb{Z}/p)^{l(I)}}\left(1 + \left\{p\ ((\mathbb{Z}/p)\times(\mathbb{Z}/p)^{l(I)})/(\mathbb{Z}/p)^{l(I)} - p)\right\}\right)_*$$

$$\subseteq \begin{cases} \bar{B} & \text{when } p=2 \\ \widetilde{\mathfrak{P}}(I_1) & \text{when } p=\text{odd}\ , \end{cases}$$

$$\mathrm{Im}\ \alpha_{(\mathbb{Z}/p)\times(\mathbb{Z}/p)^{l(I)}}\left(\{1+s\}^p\right)_* \in \widetilde{\mathfrak{P}}(I_1).$$

Here the first and the second inclusions are nothing but Lemma 3.10, and the forth was explained before; so we only have to explain about the third inclusion. When p is odd, we can apply Corollary 2.10 (ii) (as usual we used Lemma 3.11 (ii) and Lemma 1.1) amd find out the left hand side is contained in $\widetilde{\mathfrak{P}}(I_1)$. When $p=2$ and $l(I)=2$, we can't apply Corollary 2.10 (ii), but we use the additive case of Lemma 1.1 and Lemma 3.11 (ii) instead and find out the left hand side is contained in the submodule generated by $x_{(o,a)}$; so contained in $\bar{B}$. These observations complete the proof. □

Corollary 3.15. *Let* $I=(J,K)$, $l(K)\geq 2$. *Then*

$$\tilde{Q}_J(x_K) \equiv x_I + \sum b_L x_L,$$

where the congruence is taken with respect to

$$\text{mod.}\begin{cases} \bar{B}\#\bar{B} & \text{when } p=2 \\ \widetilde{\mathfrak{P}}(I_1)+\widetilde{\mathfrak{P}}(I_1)\#\widetilde{\mathfrak{P}}(I_1)+\bar{B}\#\widetilde{\mathfrak{P}}(I_1)+\bar{B}\#\bar{B} & \text{when } p=\text{odd}, \end{cases}$$

L runs over the admissible sequence with $l(K)\leq l(L)<l(I)$ *and furthermore over* $(0,a)$

$a \in \mathbb{N}$ when $p=2$ and $l(K)=2$, and b_L is some element in $\mathbb{F}_p$.

Remark 3.16. The weak form of Theorem 3.14, where the congruence is taken modulo $I_1 \# I_1$, was proved by Madsen [12] when $p=2$. But the corresponding odd primary version due to Tsuchiya [26], whose congruence is also modulo $I_1 \# I_1$, caused a question by May [5, footnote of p.141]. But, Tsuchiya's claim isn't strong enough to prove Theorem 3.21, i.e. the congruence modulo $I_1 \# I_1$ statement isn't sufficient. Maybe we should mention that the original motivation of Madsen and Tsuchiya was to prove the following (especially Corollary 3.18).

Theorem 3.17. (*i*) (*Madsen* [12]) *For $p=2$,*

$$H_*(SG) \cong E[x_1, x_2, \ldots] \otimes \mathbb{Z}/2[x_{(0,a)} \mid a>0] \otimes$$
$$\mathbb{Z}/2[\widetilde{Q}_J x_I \mid l(I)=2.\ (J,I) \text{ admissible, where } J \text{ could be empty}].$$

(*ii*) (*cf. Tsuchiya* [26], *May* [5, 6.2.]) *For $p=$ odd,*

$$H_*(SG)$$
$$\cong A[x_i \mid i \text{ admissible}] \otimes A[\widetilde{Q}_J x_I \mid l(I)=2, (J.I) \text{ admissible where } J \text{ could be empty}].$$

Corollary 3.18. *Let σ_*: $H_*(SG) \to H_{*+1}(BSG)$ be the suspension homomorphism.*

(*i*) (*Madsen* [12]) *For $p=2$,*

$$H_*(BSG) \cong H_*(BSO) \otimes E[\sigma_* x_{(0,a)} \mid a>0] \otimes \mathbb{Z}/2[x_I \mid l(I)=2.\ I \text{ admissible}] \otimes$$
$$\mathbb{Z}/2[\widetilde{Q}_{\bar{J}} \sigma_* x_I \mid l(I)=2,\ i_1 \geq 1,\ (J,I) \text{admissible, where } \bar{J}=(i_1,\cdots,i_l),\ J=(i_1+1,\cdots,i_l+1)]$$

(*ii*) (*Tsuchiya* [26]) *For $p=$ odd,*

$$H_*(BSG) \cong A[@],$$

where @ is the set consisting of the following elements;

(1) $\widetilde{z}_i$, *which is a generator of the free commutative algebra:*
Im j_* : $H_*(BSO) \to H_*(BSG)$ *with* deg $\widetilde{z}_i = 2i(p-1)$.

(2) $\sigma_* x_i$; *i admissible.*

(3) $\sigma_* x_I$; $l(I)=2$, I *admissible.*

(4) $\widetilde{Q}_{\bar{J}}\sigma_* x_I$; $l(I)=2$, $i_1 \geq 1$, (J,I) *admissible, where* $\bar{J}=(i_1,\cdots,i_l)$ *and* $J=(i_1+1,\cdots,i_l+1)$.

Now we continue our applications.

Theorem 3.19 . (*i*) *Suppose* $p=odd$, *then for* $2\leq a\leq p$ *we have*

$$x_{i_1}*\cdots*x_{i_a}*[1-a] \equiv x_{i_1}\cdots x_{i_a} + \sum_{l(I)=a} \mu_I x_I$$

where the congruence is taken with respect to

$$\text{modulo } \left(\bar{B}+\widetilde{\mathfrak{P}}(I_1)\right)\# I_1+\widetilde{\mathfrak{P}}(I_1),$$

I runs over admissible sequences with $l(I)=a$, *and* $\mu_I \in \mathbb{Z}/p$.

In fact, we can teke these so that

$$\alpha_{(\mathbb{Z}/p)^a}\left(1+ ((\mathbb{Z}/p)^a-p^a)\right)_*(e_{i_1}\otimes\cdots\otimes e_{i_a}) = \sum \mu_I x_I + \widetilde{\mathfrak{P}}(y)$$

for some $y\in I_1$.

(*ii*) (*Priddy* [21, 2.8]) *Suppose* $p=2$, *then we have*

$$x_i*x_j*[-1] \equiv x_i x_j + \sum_{l(I)=2} \mu_I x_I \ \text{mod. } \bar{B}\# I_1,$$

for some $\mu_I \in \mathbb{Z}/2$
In fact, we can take

$$\sum_{l(I)=2} \mu_I x_I = \sum_{\mathrm{Max}(0,\frac{1}{2}(j-i))\leq k\leq \frac{j}{2}} \binom{j-k}{k} x_{(i-j+2k,j-k)},$$

which can be further reduced using the Adem relation (this last remark was not noticed in Priddy [21]).

Proof (*i*). From the additive case of Lemma 1. 1,

$$x_{i_1}*\cdots*x_{i_a}*[1-a] = (Q_{i_1}[1]*[-p])*\cdots*(Q_{i_a}[1]*[-p])*[1]$$

is the homology image of $e_{i_1}\otimes\cdots\otimes e_{i_a}\in H_*(B(\mathbb{Z}/p)^a)$ by

$$B(\mathbb{Z}/p)^a=\Big(B(\mathbb{Z}/p)\Big)^a\xrightarrow{\left(\alpha_{\mathbb{Z}/p}(\mathbb{Z}/p-p)\right)^a}\Big(Q_0S^0\Big)^a\xrightarrow{\theta}Q_0S^0.$$

In the notation of Proposition 2. 15, this can be restated as

$$x_{i_1}*\cdots*x_{i_a}*[1-a]=\alpha_G\Big(1+\sum_{1\leq j\leq a}(G/H_j-p)\Big)_*(e_{i_1}\otimes\cdots\otimes e_{i_a}).$$

Of course, our philosophy was to consider the Burnside ring elemnt first. In this case, it was shown in Proposition 2. 15 (i) that

$$\begin{aligned}
&1+\sum_{1\leq j\leq a}(G/H_j-p)\\
&=\Big\{\prod_{1\leq j\leq a}\{1+(G/H_j-p)\}\Big\}\\
&\quad\times\prod_{k=2}^{a-1}\Bigg(\prod_{1\leq j_1<\dots<j_k\leq a}\{1+(G/H_{j_1j_2,\dots,j_k}-p^k)\}\Bigg)^{\alpha(k)}\\
&\quad\times\{1+(G-p^a)\}\times(1+v)^p
\end{aligned}$$

for some $v\in J=I(G)\hat{}_p$, where $\alpha(k)$ is chosen so that $1\leq\alpha(k)\leq p-1$ and $\alpha(k)\equiv(-1)^{k+1}(k-1)!\ (\text{mod}.p)$.

So, as the proof of Theorem 3. 14, we will apply Lemma 3. 11 (ii) (again thanks to the multiplicative case of Lemma 1. 1), but this time we should be careful about our particular choise of the product decomposition. Actually, we consider this as a product of the following:

<u>single</u> $\Big\{\prod_{1\leq j\leq a}\{1+(G/H_j-p)\}\Big\}$, several $\{1+(G/H_{j_1j_2,\dots,j_k}-p^k)\}$'s,

$\{1+(G-p^a)\}$, and $(1+v)^p$.

Then the relevant homology elements to apply Lemma 3. 11 are

$$\alpha_G\Big(\prod_{1\le j\le a}\{1+(G/H_j-p)\}\Big)_*(e_{i_1}\otimes\cdots\otimes e_{i_a}) = x_{i_1}\cdots x_{i_a}\ ,$$

$$\alpha_G\Big(1+(G/H_{j_1j_2,..,j_k}-p^k)\Big)_*(e_{i_1}\otimes\cdots\otimes e_{i_a}) = 0$$

for $2\le k\le a\text{-}1$ and $1\le j_1<\ldots<j_k\le a$,

$$\alpha_G\Big(1+(G-p^a)\Big)_*(e_{i_1}\otimes\cdots\otimes e_{i_a}) = \sum \mu_I x_I + \widetilde{\mathfrak{P}}(y)$$

where $y\in I_1$, $\mu_I\in\mathbb{Z}/p$, and x_I runs over admissible sequence with $l(I)=r-s$,

$$\alpha_G\Big((1+v)^p\Big)_*(e_{i_1}\otimes\cdots\otimes e_{i_a}) \in \widetilde{\mathfrak{P}}\Big(I_1\Big).$$

Here every equality follows easily as before (see the proof of Theorem 3.14), except the second equality. But this follows immediately from the factorization

$$\alpha_G\Big(1+(G/H_{j_1j_2,..,j_k}-p^k)\Big) : BG \to B(G/H_{j_1j_2,..,j_k}) \to SG\ ,$$

of which the first map sends $e_{i_1}\otimes\cdots\otimes e_{i_a}$ to 0.

Of course, our proof is still imcomplete until we deal with the modulus in the formula of Lemma 3.11(ii); the necessarilly recepie for this is:

For $*>0$,

$$\mathbf{Im}\ \alpha_G\Big(\prod_{1\le j\le a}\{1+(G/H_j-p)\}\Big)_* \subseteq I_1\ ,$$

$$\mathbf{Im}\ \alpha_G\Big(1+(G/H_{j_1j_2,..,j_k}-p^k)\Big)_* \subseteq \bar{B}\#\widetilde{\mathfrak{P}}(I_1)\ ,$$

$$\textbf{Im}\ \alpha_G\Big(1+(G-p^a)\Big)_* \subseteq \bar{B}\#\widetilde{\mathfrak{P}}(I_1)\ ,$$

$$\textbf{Im}\ \alpha_G\Big((1+v)^p\Big)_* \subseteq \widetilde{\mathfrak{P}}(I_1)\ ,$$

where the second and the third follows from Lemma 3.10 and the fourth from Lemma 3.11 (ii). These observation completes the proof.

Proof (ii) The proof is almost the same as that of (i). As before, in the notation of Proposition 2.15 (ii),

$$x_i * x_j * [-1] = \alpha_G\Big(1+(G/H_1-2)+(G/H_2)\Big)_*\Big(e_i\otimes e_j\Big),$$

and Proposition 2. 15 (ii) says

$$\begin{aligned} 1 + (G/H_1-2) + (G/H_1-2) = &\Big\{\Big\{1 + (G/H_1-2)\Big\}\Big\{1 + (G/H_2-2)\Big\}\Big\} \\ &\times \Big\{1 + 2(G/H_1-2)\Big\} \times \Big\{1 + 2(G/H_2-2)\Big\} \\ &\times \Big\{1 + (G-4)\Big\} \times (1+l)^2 \end{aligned}$$

for some $l\in J=D(G)\hat{_2}+2I(G)\hat{_2}$.

Then, as in (i), the following recepie completes the proof:

$$\alpha_G\Big(\Big\{1 + (G/H_1-2)\Big\}\Big\{1 + (G/H_2-2)\Big\}\Big)_*\Big(e_i\otimes e_j\Big) = x_i x_j,$$

$$\alpha_G\Big(\Big\{1 + 2(G/H_1-2)\Big\}\Big)_*\Big(e_i\otimes e_j\Big) = 0,$$

$$\alpha_G\Big(1 + 2(G/H_2-2)\Big)_*\Big(e_i\otimes e_j\Big) = 0,$$

$$\alpha_G\Big(1 + (G-4)\Big)_*\Big(e_i\otimes e_j\Big) = \sum_{\mathrm{Max}(0,\frac{1}{2}(j-i))\le k\le\frac{j}{2}} \binom{j-k}{k} x_{(i-j+2k,j-k)} \quad \cdots (1)$$

$\alpha_G\left((1+I)^2\right)_*\left(e_i\otimes e_j\right) \in \widetilde{\mathfrak{P}}(I_1) \subset \bar{B}\#I_1$ (this latter inclusion is only true for $p=2$).
And for $*>0$,

$$\left.\begin{aligned}
&\mathrm{Im}\ \alpha_G\left(\left\{1 + (G/H_1-2)\right\}\left\{1 + (G/H_2-2)\right\}\right)_* \subseteq I_1\,, \\
&\mathrm{Im}\ \alpha_G\left(\left\{1 + 2(G/H_1-2)\right\}\right)_* \subseteq \bar{B}, \\
&\mathrm{Im}\ \alpha_G\left(1 + 2(G/H_2-2)\right)_* \subseteq \bar{B}, \\
&\mathrm{Im}\ \alpha_G\left(1 + (G-4)\right)_* \subseteq \bar{B}, \\
&\mathrm{Im}\ \alpha_G\left((1+I)^2\right)_* \subseteq \widetilde{\mathfrak{P}}(I_1) \subset \bar{B} \quad \text{(this latter inclusion is only true for } p=2).
\end{aligned}\right\}\cdots (2)$$

Here most of the equalities and inclusions follow exactly as in the proofs of (i) and Theorem 3.14, except (1) and (2).

For (1), we note

$$\alpha_G\left(1 + (G-4)\right)_*\left(e_i\otimes e_j\right)$$

$$=\alpha_{\Sigma_4}\left(1 + (\Sigma_4/\Sigma_3-4)\right)_*\circ\, i_{\mathfrak{I}(4,2)\to\Sigma_4}\circ\, i_{(\mathbb{Z}/p)^k\to\mathfrak{I}(p^k,p)}\left(e_i\otimes e_j\right),$$

where the notations of subgroups are qw were given at the beginning of §2.
Now, by [29],

$$i_{(\mathbb{Z}/p)^k\to\mathfrak{I}(p^k,p)}\left(e_i\otimes e_j\right)$$

$$=\sum_k e_{i-j+2k}\textstyle\int(Sq_*^k e_j) = \displaystyle\sum_{\mathrm{Max}(0,\frac{1}{2}(j-i))\le k\le\frac{j}{2}} \binom{j-k}{k}\, e_{i-j+2k}\textstyle\int e_{j-k}\,.$$

So (1) now follows from Lemma 3.9.

For (2), note

$$\alpha_G\Big(1 + 2(G/H_n-2)\Big) = \alpha_G\Big(2(G/H_n-2)\Big)_* *[1] \qquad (n=1 \text{ or } 2).$$

Then, by Lemma 3.11(ii), the claim follows from

$$Q_0\Big(Q_a[1]*[-2]\Big)*[1] = Q_0 Q_a[1]*[-3] = x_{(0,a)} \in \overline{B}. \qquad \square$$

Theorem 3.20. *Suppose p=odd, then we have*

$$\mathfrak{P}(x_i*[-1])*[1] \equiv \widetilde{\mathfrak{P}}(x_i) \text{ mod. } \widetilde{\mathfrak{P}}^2(I_1)+\widetilde{\mathfrak{P}}(I_1)\#\widetilde{\mathfrak{P}}^2(I_1).$$

Proof. If i is odd, then from Theorem 3.7 and Theorem 3.8.(*ii*)

$$\mathfrak{P}(x_i*[-1])*[1] = \widetilde{\mathfrak{P}}(x_i) = 0.$$

So we may suppose i is even (actually $i\equiv 0$ mod. $2(p-1)$). Then from the structure of $H^*(B\mathbb{Z}/p)$, we can find some $f \in H_*(B\mathbb{Z}/p)$ such that

$$\Delta_*(f) = e_i\otimes\cdots\otimes e_i + \sum_{(\lambda_1\lambda_2\cdots\lambda_p)}\left\{\sum_{(i_1i_2\cdots i_p)\in Cycl.(\lambda_1\lambda_2\cdots\lambda_p)} (-1)^{?} g_{i_1}\otimes g_{i_2}\otimes\cdots\otimes g_{i_p}\right\},$$

where $Cycl.(\lambda_1\lambda_2\cdots\lambda_p)$ is the set of the cyclic permutations of $(\lambda_1\lambda_2\cdots\lambda_p)$. (So $\#|Cycl.(\lambda_1\lambda_2\cdots\lambda_p)| = p$.) Note that for any map m from $B\mathbb{Z}/p$ to a homotopy commutative and associative H-space X,

$$\sum_{(i_1i_2\cdots i_p)\in Cycl.(\lambda_1\lambda_2\cdots\lambda_p)} (-1)^{?} m_*(g_{i_1})\otimes m_*(g_{i_2})\otimes\cdots\otimes m_*(g_{i_p}) = 0 \in H_*(X).$$

Therefore

$$\mathfrak{P}(x_i*[-1])*[1] = \alpha_{B\mathbb{Z}/p}(1+p(\mathbb{Z}/p - p))_*(f). \qquad \cdots (*)$$

According to our philosophy, we should study $1+p(\mathbb{Z}/p - p)$ first. We are going to take somewhat indirect approach to this.

First, using

$$(\mathbb{Z}/p-p)^2 = -p(\mathbb{Z}/p-p),$$

we get

$$(1+(\mathbb{Z}/p-p))^p = 1+\frac{(1-p)^p-1}{-p}(\mathbb{Z}/p-p).$$

Since

$$\frac{(1-p)^p-1}{-p} = p+p^2n$$

for some $n\in\mathbb{N}$, we can write

$$(1+(\mathbb{Z}/p-p))^p$$

$$= 1+(p+p^2n)(\mathbb{Z}/p-p) = \left\{1+p(\mathbb{Z}/p-p)\right\}\left\{1+p^2m(\mathbb{Z}/p-p)\right\},$$

with appropriate $m\in\mathbb{N}$.

Senond. at this stage, we have to deal with $1+p^2m(\mathbb{Z}/p-p)$, but as we can easily see

$$log_p\left\{1+p^2m'(\mathbb{Z}/p-p)\right\} \in p^2I(\mathbb{Z}/p)\hat{_p},$$

we can put

$$1+p^2m'(\mathbb{Z}/p-p) = exp_p\circ log_p\left\{1+p^2m'(\mathbb{Z}/p-p)\right\} = (1+d)^{p^2},$$

with some $d\in I(\mathbb{Z}/p)\hat{_p}$ (see Theorem 2.9).

Then we get our sought after expression

$$1+p(\mathbb{Z}/p-p) = (1+(\mathbb{Z}/p-p))^p(1+d')^{p^2},$$

with some $d'\in\mathbb{N}$. Now the theorem follows from $(*)$ and Lemma 3.11 as before (see the proofs of Theorem 3.14 and Theorem 3.19). □

Now we go to the multiplicative transfer theorems, which are strongly motivated by Corollary 3.15 and could be regarded as the multiplicative analogue of the famous Kahn-Priddy theorem [10].

Theorem 3.21. (*Priddy* [21] *for* $p=2$) *Let* δ *be the composite:*

$$B\mathbb{Z}_p \int \mathbb{Z}_p \to B\Sigma_{p^2} \xrightarrow{D\text{-}L} Q_{p^2}S^0 \xrightarrow{*[1-p^2]} Q_1S^0 = SG.$$

Then there is a map $t\colon SG \to Q(B\mathbb{Z}_p \int \mathbb{Z}_p)$ *such that*

$$SG \xrightarrow{t} Q(B\mathbb{Z}_p \int \mathbb{Z}_p) \xrightarrow{Q(\delta)} QSG \xrightarrow{\xi} SG$$

is an equivalence at p, *where* ξ *is the map induced by the infinite loop space structure of* SG.

Corollary 3.22. *We suppose* $p=odd$. *Let* γ *be the composite:*

$$B\mathbb{Z}_p \int \mathbb{Z}_p \to B\Sigma_{p^2} \xrightarrow{D\text{-}L} Q_{p^2}S^0 \xrightarrow{*[1-p^2]} Q_1S^0 = SG \to C_\otimes,$$

where the last map is the infinite loop space splitting of SG *to the multiplicative* $CokerJ$ *space (This is given by the theorem of Tornhave* [24]; *see also* [5]). *Then there is a map* $u\colon C_\otimes \to Q(B\mathbb{Z}_p \int \mathbb{Z}_p)$ *such that*

$$C_\otimes \xrightarrow{u} Q(B\mathbb{Z}_p \int \mathbb{Z}_p) \xrightarrow{Q(\gamma)} QC_\otimes \xrightarrow{\xi'} C_\otimes$$

is an equivalence at p, *where* ξ' *is the map induced by the infinite loop space structure of* $C_\otimes$.

In [21], δ is denoted by δ_1. Now we begin our proof. We follow the method of [21][10], which uses the homology operations. Concerning the case $p=2$, we note that our proof (filtration used) is much less complicated than Priddy's original proof [21] since our Theorem 3.14 is sharper than Madsen's original result [12].

Proof of Theorem 3.21. By the standard argument (see [21], [10]), we only have to show the composite

$$C\colon \Sigma^\infty B\Sigma_{p^k} \xrightarrow{\tau} \Sigma^\infty B\mathcal{S}(p^k,p) \xrightarrow{\Sigma^\infty Bi} \Sigma^\infty B(\Sigma_{p^{k-2}}\int \mathbb{Z}/p\int \mathbb{Z}/p)$$

$$\simeq \Sigma^\infty E\Sigma_{p^{k-2}} \underset{\Sigma_{p^{k-2}}}{\times} (B(\mathbb{Z}/p\int \mathbb{Z}/p))^{p^{k-2}} \xrightarrow{\Sigma^\infty(D\text{-}L)} \Sigma^\infty Q(B(\mathbb{Z}/p\int \mathbb{Z}/p))$$

$$\xrightarrow{\Sigma^\infty Q\delta} \Sigma^\infty QSG \xrightarrow{\Sigma^\infty \xi} \Sigma^\infty SG$$

induces an isomorphism of mod-p homology in a range of dimensions which increases along with some sequence of k increases, where τ is the stable transfer associated with the finite covering and induces $tr_{\Sigma_{p^k}\to \mathcal{S}(p^k,p)}$, and where $i\colon \mathcal{S}(p^k,p) \to \Sigma_{p^{k-2}}\int \mathbb{Z}/p\int \mathbb{Z}/p$ is the inclusion.

To show this, we define a decreasing filtration in $H_*(SG)$. Pick a monomial

$$x_{i_1}^{pa_1+b_1} x_{i_2}^{pa_2+b_2} \cdots x_{i_m}^{pa_m+b_m} x_{I_1} x_{I_2} \cdots x_{I_n}$$

such that

1) i_j admissible, $a_j \geq 0$, $0 \leq b_j < p$, and $i_j \neq i_k$, for $j \neq k$ and $1 \leq j,k \leq m$.

2) $l(I_l) \geq 2$, and I_l admissible or (when $p=2$) $=$ (O,a) for some a, for $1 \leq l \leq n$.

3) when $p=2$, of course we may suppose $a_j=0$ and $i_1 < i_2 < \cdots < i_m$.

We then define

$$f(x_{i_1}^{pa_1+b_1} x_{i_2}^{pa_2+b_2} \cdots x_{i_m}^{pa_m+b_m} x_{I_1} x_{I_2} \cdots x_{I_n}) = \sum_{1\leq j\leq m} (p^2 a_j + b_j) + pn.$$

This is extended from the above $\mathbb{Z}/p$-base to the whole $H_*(SG)$ as usual by setting

$$f\left(\sum_j \lambda_j x_{i_{j,1}}^{pa_{j,1}+b_{j,1}} x_{i_{j,2}}^{pa_{j,2}+b_{j,2}} \cdots x_{i_{j,m_j}}^{pa_{j,m_j}+b_{j,m_j}} x_{I_{j;1}} x_{I_{j,2}} \cdots x_{I_{j,n_j}}\right)$$

$$= \operatorname*{Min}_j f(x_{i_{j,1}}^{pa_{j,1}+b_{j,1}} x_{i_{j,2}}^{pa_{j,2}+b_{j,2}} \cdots x_{i_{j,m_j}}^{pa_{j,m_j}+b_{j,m_j}} x_{I_{j,1}} x_{I_{j,2}} \cdots x_{I_{j,n_j}}),$$

where $\lambda_j \in \mathbb{Z}/p-\{0\}$. From this, we set

$$F_l = \{u \in H_*(SG) \mid l \leq f(u)\}.$$

Obviously $H_*(SG) = F_0 \supseteq F_1 \supseteq \cdots \supseteq F_l \supseteq \cdots$, $F_a F_b \subseteq F_{a+b}$, and $F_z = 0$ for some z which is depending upon just the degree $*$. One of the properties of this filtration we should remember is the following:

$$\textit{For any } y \in \widetilde{\mathfrak{P}}(I_1),\ y \in F_{p^2}. \tag{$\bigstar$}$$

We then define an increasing filtration by setting

$$g\Big(\sum_j \lambda_j x_{i_{j,1}} x_{i_{j,2}} \cdots x_{i_{j,m_j}} x_{I_{j,1}} x_{I_{j,2}} \cdots x_{I_{j,n_j}}\Big) = \operatorname*{Max}_j \Big(pm_j + \sum_{1 \leq k \leq n_j} l(I_{j,k})\Big)$$

where $\lambda_j \in \mathbb{Z}/p-\{0\}$, and we put

$$G_m = \{v \in H_*(SG) \mid m \geq g(v)\}.$$

Now the filtration we are going to use is the following

$$F_{l,m} = \text{sub } \mathbb{Z}/p\text{-module generated by } \{F_l \cap G_m\} \cup F_{l+1}.$$

From the definition, for any l there exists some m_l such that $F_{l,m_l} = F_l$ (we note that we are fixing the degree $*$). Using this, we immediately see

$$\begin{aligned}
H_*(SG) &= F_0 = F_{0,m_0} \supseteq F_{0,m_0-1} \supseteq F_{0,m_0-2} \supseteq \cdots \supseteq F_{0,0} \\
&= F_1 = F_{1,m_1} \supseteq F_{1,m_1-1} \supseteq F_{1,m_1-2} \supseteq \cdots \supseteq F_{1,0} \\
&= F_2 = F_{2,m_2} \supseteq F_{2,m_2-1} \supseteq F_{2,m_2-2} \supseteq \cdots \supseteq F_{2,0} \\
&= F_3 = F_{3,m_3} \supseteq \cdots \\
&\supseteq \cdots\cdots \supseteq F_{l-1,0} \\
&= F_l = F_{l,m_l} \supseteq F_{l,m_l-1} \supseteq F_{l,m_l-2} \supseteq \cdots \supseteq F_{l,0} \\
&= F_{l+1} = F_{l+1,m_{l+1}} \supseteq \cdots \\
&\supseteq \cdots\cdots \supseteq F_{z-2,0} \\
&= F_{z-1} = F_{z-1,m_{z-1}} \supseteq F_{z-1,m_{z-1}-1} \supseteq F_{z-1,m_{z-1}-2} \supseteq \cdots \supseteq F_{z-1,0} \\
&= F_z = 0,
\end{aligned}$$

and $F_{a,b}F_{c,d} \subseteq F_{a+c,b+d}$. We are going to show, for any $x \in F_{l,m}$, there exists some $y \in H_*(B\Sigma_{p^k})$ such that

$$C_*(y) \equiv x \ \text{mod.}\ F_{l,m-1}.$$

Of course, this would complete the proof from the above remark. At first we restate some of our previous results in terms of the filtration just mentioned. We remind the reader the important ($\bigstar$) $\widetilde{\mathfrak{P}}(I_1) \subseteq F_{p^2}$. Then theorems 3.13, 3.18 and 3.19 can be restated as follows:

(1) *Let p be any prime and let (J,I) be any non-negative sequence such that $l(I) \geq 2$, then*

$$\widetilde{Q}_J(x_I) \equiv x_{(J,I)} \ \text{mod.} F_{p,\ (l(J)+l(I)-1)}.$$

$$\textit{whereas } x_{(J,I)} \in F_{p,\ (l(J)+l(I))} \supseteq F_{p,\ (l(J)+l(I)-1)}.$$

(2) *Suppose p is odd and $2 \leq a < p$, then*

$$x_{i_1} * \cdots * x_{i_a} * [1-a] \equiv x_{i_1} \cdots x_{i_a} \quad \text{mod.}\ F_{p,a}.$$

$$\textit{whereas } x_{i_1} \cdots x_{i_a} \in F_a \supseteq F_{p,a}.$$

(3) *Let p be any prime. If $a=p$ and $i_j \neq i_k$ for some $1 \leq j,\ k \leq p$, then*

$$x_{i_1} * \cdots * x_{i_p} * [1-p] \equiv x_{i_1} \cdots x_{i_p} \quad \text{mod.}\ F_{p,p},$$

$$\textit{whereas } x_{i_1} \cdots x_{i_p} \in F_{p,p2} \supseteq F_{p,p}.$$

(4) *When p is odd, we have*

$$x_i * \cdots * x_i = x_{(0,i)} \equiv x_i^{\ p} \quad \text{mod.}\ F_{p^3}$$

$$\textit{whereas } x_i^{\ p} \in F_{p^2} \supseteq F_{p^3}.$$

If I is admissible with $l(I) \geq 2$, it's easy to see $x_I \in F_{p,pl(I)}$. Now (1) and (3) show that this generally holds for arbitrary non-negative sequence:

(5) *For any sequence I with $l(I) \geq 2$, we have $x_I \in F_{p,l(I)}$.*

Proof of (5). From the Adem relation (cf. 3.4 and 3.8), we only have to check when I is weakly admissible and not admissible. By Proposition 3.11, this x_I is either contained in $\widetilde{\mathfrak{P}}(I_1)$ or of the form $(0,a)$. The latter case is of course O.K. The former case follows from $(\bigstar)$.

Now let's begin. Pick up an element of monomial base in $H_*(SG)$ say

$$x = x_{i_1}^{pa_1+b_1} x_{i_2}^{pa_2+b_2} \cdots x_{i_m}^{pa_m+b_m} x_{I_1} x_{I_2} \cdots x_{I_l} x_{I_{l+1}} x_{I_{l+2}} \cdots x_{I_{l+n}}$$

such that

1) i_j admissible, $a_j \geq 0$, $0 \leq b_j < p$, and $i_j \neq i_k$, for $j \neq k$ and $1 \leq j,k \leq m$.

2) $l(I_s)=2$ for $1 \leq s \leq l$, $l(I_t)>2$ for $l+1 \leq t \leq l+n$, and I_u admissible or (when $p=2$) $=(0,a)$ for $1 \leq u \leq l+n$.

3) when $p=2$, of course we may suppose $a_j=0$ and $i_1 < i_2 < \cdots < i_m$.

Of course $x \in F_{\sigma,\tau}$ where

$$\sigma = \sum_{1 \leq k \leq m} (p^2 a_k + b_k) + p(l+m) \text{ and } \tau = p \sum_{1 \leq k \leq m} (pa_k + b_k) + \sum_{1 \leq j \leq l+m} l(I_j).$$

For our purpose, it is convenient to rewrite x as follows:

$$x = x_{j_1}^{p} x_{j_2}^{p} \cdots x_{j_r}^{p} x_{k_1} \cdots x_{k_s} x_{I_1} x_{I_2} \cdots x_{I_l} x_{I_{l+1}} x_{I_{l+2}} \cdots x_{I_{l+m}},$$

where

$$x_{j_1}^{p} x_{j_2}^{p} \cdots x_{j_r}^{p} = x_{i_1}^{pa_1} x_{i_2}^{pa_2} \cdots x_{i_m}^{pa_m}$$

and

$$x_{k_1} \cdots x_{k_s} = x_{i_1}^{b_1} x_{i_2}^{b_2} \cdots x_{i_m}^{b_m}.$$

So every x_{j_i} and x_{k_j} is one of x_{i_k} $(1 \leq k \leq m)$ and $r = \sum_{1 \leq k \leq m} a_k$, $s = \sum_{1 \leq k \leq m} b_k$. Now we only have to find some $e \in H_*(B\Sigma_{p^k})$ such that

$$C_*(e) \equiv x \quad \text{mod. } F_{\sigma,\tau-1}$$

when k is even and $p^k \geq p \sum_{1 \leq k \leq m} (pa_k + b_k) + \sum_{1 \leq j \leq l+m} p^{l(I_j)}$. We now claim the element

$$
\begin{aligned}
&e \\
&= \mathfrak{P}(e_{j_1})*\mathfrak{P}(e_{j_2})*\cdots*\mathfrak{P}(e_{j_r})*e_{k_1}*\cdots*e_{k_s}*e_{I_1}*e_{I_2}*\cdots*e_{I_l}*e_{I_{l+1}}*\cdots*e_{I_{l+m}} \\
&\qquad =(e_{j_1}*\cdots*e_{j_1})*\cdots*(e_{j_r}*\cdots*e_{j_r})*e_{k_1}*\cdots*e_{k_s}*e_{I_1}*\cdots*e_{I_l}*e_{I_{l+1}}*\cdots*e_{I_{l+m}}
\end{aligned}
$$

is a sought after one. In fact, putting $D=(\Sigma^{\infty}\xi)\circ(\Sigma^{\infty}Q\delta)\circ(\Sigma^{\infty}D\text{-}L)\circ(\Sigma^{\infty}Bi)$, we have

$$
\begin{aligned}
&C_*(e) = D\circ\tau(e) \\
&= D\Big(\hat{e}_{j_1}|\cdots|\hat{e}_{j_1}|\cdots|\hat{e}_{j_r}|\cdots|\hat{e}_{j_r}|\hat{e}_{k_1}|\cdots|\hat{e}_{k_s}|\hat{e}_{I_1}|\hat{e}_{I_2}|\cdots|\hat{e}_{I_l}|\hat{e}_{I_{l+1}}|\cdots|\hat{e}_{I_{l+m}} \\
&+\sum \hat{e}_{j_1}|\cdots|\hat{e}_{j_1}|\cdots|\hat{e}_{j_r}|\cdots|\hat{e}_{j_r}|\hat{e}_{k_1}|\cdots|\hat{e}_{k_s}|\hat{e}_{I'_1}|\hat{e}_{I'_2}|\cdots|\hat{e}_{I'_l}|\hat{e}_{I'_{l+1}}|\cdots|\hat{e}_{I'_{l+m}}\Big) \\
&= (x_{j_1}*\cdots*x_{j_1}*[1-p])\cdots(x_{j_r}*\cdots*x_{j_r}*[1-p])(x_{k_1}*\cdots*x_{k_p}*[1-p])\cdots(\cdots*x_{k_s}*[1-?]) \\
&\times x_{I_1}x_{I_2}\cdots x_{I_l}\times\widetilde{Q}_{J_{l+1}}(x_{K_{l+1}})\widetilde{Q}_{J_{l+1}}(x_{K_{l+1}})\cdots\widetilde{Q}_{J_{l+m}}(x_{K_{l+m}}) \\
&+\sum(x_{j_1}*\cdots*x_{j_1}*[1-p])\cdots(x_{j_r}*\cdots*x_{j_r}*[1-p])(x_{k_1}*\cdots*x_{k\ p}*[1-p])\cdots(\cdots*x_{k_s}*[1-?]) \\
&\times x_{I'_1}x_{I'_2}\cdots x_{I'_l}\times\widetilde{Q}_{J'_{l+1}}(x_{K'_{l+1}})\widetilde{Q}_{J'_{l+1}}(x_{K'_{l+1}})\cdots\widetilde{Q}_{J'_{l+m}}(x_{K'_{l+m}}) \\
&\equiv x_{j_1}{}^{p}x_{j_2}{}^{p}\cdots x_{j_r}{}^{p}x_{k_1}\cdots x_{k_s}\times x_{I_1}x_{I_2}\cdots x_{I_l}\times x_{I_{l+1}}\cdots x_{I_{l+m}} \\
&+\sum x_{j_1}{}^{p}x_{j_2}{}^{p}\cdots x_{j_r}{}^{p}x_{k_1}\cdots x_{k_s}\times x_{I'_1}x_{I'_2}\cdots x_{I'_l}\times x_{I'_{l+1}}\cdots x_{I'_{l+m}},
\end{aligned}
$$

where the congruence is mod. $F_{\sigma,\tau-1}$, guaranteed by the above (1), (2), (3), (4) and (5) (together with the multiplicative property of our filtration, $F_{a,b}\ F_{c,d}\subseteq F_{a+c,b+d}$).

To prove

$$
C_*(e)\equiv x\left(=x_{j_1}{}^{p}x_{j_2}{}^{p}\cdots x_{j_r}{}^{p}x_{k_1}\cdots x_{k_s}\times x_{I_1}x_{I_2}\cdots x_{I_l}\times x_{I_{l+1}}\cdots x_{I_{l+m}}\right)
$$

$$
\text{mod. } F_{\sigma,\tau-1},
$$

we may show (when p is odd, we assume k is even)

$$\sum x_{I'_1} x_{I'_2} \cdots x_{I'_l} x_{I'_{l+1}} \cdots x_{I'_{l+m}} \equiv 0 \mod. F_{p(l+m), \sum_{1 \le i \le l+m} l(I'_i) - 1},$$

because

$$\begin{aligned} &\sum x_{j_1}{}^p x_{j_2}{}^p \cdots x_{j_r}{}^p x_{k_1} \cdots x_{k_s} \times x_{I'_1} x_{I'_2} \cdots x_{I'_l} \times x_{I'_{l+1}} \cdots x_{I'_{l+m}} \\ &= \left(x_{j_1}{}^p x_{j_2}{}^p \cdots x_{j_r}{}^p x_{k_1} \cdots x_{k_s}\right) \times \left(\sum x_{I'_1} x_{I'_2} \cdots x_{I'_l} x_{I'_{l+1}} \cdots x_{I'_{l+m}}\right). \end{aligned}$$

By Theorem 3.5 and Remark 3.6

$$\begin{aligned} &\sum (e_{j_1} * \cdots * e_{j_1}) * \cdots * (e_{j_r} * \cdots * e_{j_r}) * e_{k_1} * \cdots * e_{k_s} * e_{I'_1} * e_{I'_2} * \cdots * e_{I'_l} * e_{I'_{l+1}} * \cdots * e_{I'_{l+m}} \\ &(= \sum \mathfrak{P}(e_{j_1}) * \mathfrak{P}(e_{j_2}) * \cdots * \mathfrak{P}(e_{j_r}) * e_{k_1} * \cdots * e_{k_s} * e_{I'_1} * e_{I'_2} * \cdots * e_{I'_l} * e_{I'_{l+1}} * \cdots * e_{I'_{l+m}}) \\ &= 0, \end{aligned}$$

for any k when $p=2$ and for even k when p is odd. From

$$\begin{aligned} &\sum (e_{j_1} * \cdots * e_{j_1}) * \cdots * (e_{j_r} * \cdots * e_{j_r}) * e_{k_1} * \cdots * e_{k_s} * e_{I'_1} * e_{I'_2} * \cdots * e_{I'_l} * e_{I'_{l+1}} * \cdots * e_{I'_{l+m}} \\ &= \left((e_{j_1} * \cdots * e_{j_1}) * \cdots * (e_{j_r} * \cdots * e_{j_r}) * e_{k_1} * \cdots * e_{k_s}\right) \\ &\qquad * \left(\sum e_{I'_1} * e_{I'_2} * \cdots * e_{I'_l} * e_{I'_{l+1}} * \cdots * e_{I'_{l+m}}\right) \end{aligned}$$

and the structure of $H_*(Q_0 S^0)$, we deduce that

$$\sum e_{I'_1} * e_{I'_2} * \cdots * e_{I'_l} * e_{I'_{l+1}} * \cdots * e_{I'_{l+m}} = 0.$$

Even though the structures of $H_*(SG)$ and $H_*(Q_0 S^0)$ are very similar, still an obstacle exists since there's no guarantee that each sequence I'_i is admissible. To proceed further we should make precise the similarity of the structures of $H_*(SG)$ and

$H_*(Q_0S^0)$. We put

$$S=\begin{cases}\mathbb{Z}/2[e_{(0,a)},\ e_I \mid a>0,\ I\ admissible] & p=2\\ H_*(Q_0S^0) & p=odd\end{cases},$$

then clearly $S\subseteq H_*(Q_0S^0)$ and there is the canonical ring isomorphism $S\cong R$ $\left(\subseteq H_*(SG)\right)$ (recall the definition of R in Proposition 3.12). Then we can define exactly the same filtration on S as the filtration $F_{l,m}$ on $R\subseteq H_*(SG)$, and we denote by $\overline{F}_{l,m}$ the corresponding filtration on S. Now, using the Adem relation (Theorem 3.4.), we can express each e_{I_i} as follows:

$$e_{I_i}=\sum e_{I''_{i,v_i}}+\mathfrak{P}(f),$$

where I''_{i,v_i} runs over admissible sequences and when $p=2$ further over elements of the form $(0,a)$, and $f\in S$. The point is the fact $S=R*[-1]$, due to Milgram [16] (in [16] R and S is denoted B and A respectively) for $p=2$ and Tsuchiya [25] and May [5] for p=odd. This enables us to apply Proposition 3.12 and deduce the following:

$$x_{I_i}=\sum x_{I''_{i,v_i}}+\widetilde{\mathfrak{P}}(y),$$

where y is some element of R and of course $x_{I''_{i,v_i}}=[1]*e_{I''_{i,v_i}}$.

Now what we should do is self-evident. First, from (★) and its corresponding version for S, we get the following:

$$\sum x_{I_1}x_{I_2}\cdots x_{I_l}x_{I_{l+1}}\cdots x_{I_{l+m}}$$

$$\equiv\sum x_{I''_{1,v_1}}x_{I''_{2,v_2}}\cdots x_{I''_{l+m,v_{l+m}}}\quad \text{mod. } F_{p(l+m),\ \sum_{1\le i\le l+m} l(I'_i)-1} \tag{1}$$

and

$$0=\sum e_{I_1}e_{I_2}\cdots e_{I_l}e_{I_{l+1}}\cdots e_{I_{l+m}}$$

$$\equiv\sum e_{I''_{1,v_1}}e_{I''_{2,v_2}}\cdots e_{I''_{l+m,v_{l+m}}}\quad \text{mod. } \overline{F}_{p(l+m),\ \sum_{1\le i\le l+m} l(I'_i)-1}. \tag{2}$$

Second, since the filtrations $F_{l,m}$ and $\overline{F}_{l,m}$ are defined exactly in the same way in terms of admissible sequences and when $p=2$ further of elements of the form $(0,a)$, the above congruence (2) implies

$$\sum x_{I''_{1,v_1}} x_{I''_{2,v_2}} \cdots x_{I''_{l+m,v_{l+m}}} \equiv 0 \mod. F_{p(l+m),\ \sum_{1\le i\le l+m} l(I'_i)-1},$$

which further implies

$$\sum x_{I_1} x_{I_2} \cdots x_{I_l} x_{I_{l+1}} \cdots x_{I_{l+m}} \equiv 0 \mod. F_{p(l+m),\ \sum_{1\le i\le l+m} l(I'_i)-1},$$

by the above congruence (1).

Thus we get

$$C_*(e) \equiv x \mod. F_{\sigma,\tau-1},$$

as was desired. □

REFERENCES

[1] J. Adem: The relations on Steenrod powers of cohomology classes. *Algebraic geometry and topology, A symposium in honor of S. Lefschetz* Princeton. 1957.

[2] S. Araki and T. Kudo: Topology of H_n-spaces and H-squaring operations. *Mem. Fac. Sci. Kyusyu Univ. Ser. A* 10 (1956), 85 -120.

[3] J. C. Becker and D. H. Gottlieb: The transfer map and fiber bundles. *Topology* 14 (1975), 1-12.

[4] J. C. Becker and R. Schultz: Equivariant function spaces and stable homotopy theory I. *Comment. Math. Helvetici* 49 (1974), 1-34.

[5] F. Cohen, T. Lada, J. P. May: The homology of iterated loop spaces. *Lecture Notes in Math.* v533 Springer-Verlag, Berlin -New York. 1976.

[6] T. tom Dieck: Transformation groups and representation theory. *Lecture Notes in Math.* v766 Springer-Verlag, Berlin-New York. 1979.

[7] E. Dyer and R. Lashof: Homology of iterated loop spaces. *Amer. J. Math.* 84 (1962), 35-58.

[8] L. Evens: A generalization of the transfer map in the cohomology of groups. *Trans. Amer. Math Soc.* 108 (1963), 54 -65.

[9] D. S. Kahn and S. B. Priddy: On the transfer in the homology of the symmetric group. *Math. Proc. Camb. Phil. Soc.* 83 (1978), 91-101.

[10] D. S. Kahn and S. B. Priddy: The transfer and the stable homotopy theory. *Math. Proc. Camb. Phil. Soc.* **83** (1978), 103 -111.

[11] L. G. Lewis, Jr., J. P. May and M. Steinberger (with contributions by J. E. McClure): Equivariant Stable Homotopy heory. *Lecture Notes in Math.* v1213 Springer-Verlag, Berlin -New York. 1986.

[12] I. Madsen: On the action of the Dyer-Lashof algebra in $H_*(G)$. *Pacific J. Math.* **60** (1975), 235-275.

[13] J. P. May: The geometry of iterated loop spaces. *Lecture Notes in Math.* v271 Springer-Verlag, Berlin-New York. 1972.

[14] J. P. May (with contributions by F. Quinn, N. Ray and J. Tornehave): E_∞ ring spaces and E_∞ ring spectra. *Lecture Notes in Math.* v577 Springer-Verlag, Berlin-New York. 1977.

[15] J. P. May: Multiplicative infinite loop space theory. *J. Pure and Applied Algebra* **26** (1982), 1-69.

[16] R. J. Milgram: The mod 2 spherical characteristic classes. *Annals Math.* **92** (1970), 238-261.

[17] M. Nakaoka: Decomposition theorem for homology groups of symmetric groups. *Annals of Math.* **71** (1960), 16-42.

[18] M. Nakaoka: Homology of the infinite symmetric group. *Annals of Math.* **73** (1961), 229-257.

[19] M. Nakaoka: Note on cohomology algebras of symmetric groups. *J. Math. Osaka City Univ.* **13** (1962), 45-55.

[20] S. Priddy: On $\Omega^\infty S^\infty$ and the infinite symmetric group. *Algebraic topology (Proc. Sympos. Pure Math., Vol. XXII, Univ. Wisconsin Madison, Wis.*, 1970) Amer. Math. Soc., Providence, R. I., 1971, 217-220.

[21] S. Priddy: Homotopy splittings involving *G and G/O.* *Comment. Math. Helvetici* **53** (1978), 470-484.

[22] F. W. Roush: Transfer in generalized cohomology theories. *Thesis* Princeton University 1971.

[23] R. Steiner: A canonical operad pair. *Math. Proc. Camb. Phil. Soc.* **86** (1979), 443-447.

[24] J. Tornehave: The splitting of spherical fibration theory at odd primes, (*unpublished*).

[25] A. Tsuchiya: Characteristic classes for spherical fiber spaces. *Nagoya Math. J.* **43** (1971), 1-39.

[26] A. Tsuchiya: Homology operations on ring spectrum of H^∞ type and their applications. *J. Math. Soc. Japan* 25 (1973), 277-316.

[27] N. Minami: Northwestern University Thesis (1988).

[28] R. E. Bruner, J. P. May, J. E. McClure and M. Steinberger: H_∞ Ring Spectra and their applications. *Lecture Notes in Math.* v1176 Springer-Verlag, Berlin-New York. 1986.

[29] N. Steenrod and D. Epstein: Cohomology Operations. *Annals of Math. Studies* **50**, Princeton Univ. Press. 1962.

Department of Mathematics
Northwestern University
Evanston, IL 60208
U.S.A.

Department of Mathematics
Faculty of Science
Hiroshima University
Hiroshima 730
Japan

Current address:

Department of Mathematics
Massachusetts Institute of Technology
Cambridge, MA 02139
U.S.A.

Mathematical Sciences Research Institute
1000 *Centennial Drive*
Berkeley, CA 94720
U.S.A.

MEMOIRS of the American Mathematical Society

SUBMISSION. This journal is designed particularly for long research papers (and groups of cognate papers) in pure and applied mathematics. The papers, in general, are longer than those in the TRANSACTIONS of the American Mathematical Society, with which it shares an editorial committee. Mathematical papers intended for publication in the Memoirs should be addressed to one of the editors:

Ordinary differential equations, partial differential equations and applied mathematics to ROGER D. NUSSBAUM, Department of Mathematics, Rutgers University, New Brunswick, NJ 08903

Harmonic analysis, representation theory and Lie theory to AVNER D. ASH, Department of Mathematics, The Ohio State University, 231 West 18th Avenue, Columbus, OH 43210

Abstract analysis to MASAMICHI TAKESAKI, Department of Mathematics, University of California, Los Angeles, CA 90024

Real and harmonic analysis to DAVID JERISON, Department of Mathematics, M.I.T., Rm 2–180, Cambridge, MA 02139

Algebra and algebraic geometry to JUDITH D. SALLY, Department of Mathematics, Northwestern University, Evanston, IL 60208

Geometric topology and general topology to JAMES W. CANNON, Department of Mathematics, Brigham Young University, Provo, UT 84602

Algebraic topology and differential topology to RALPH COHEN, Department of Mathematics, Stanford University, Stanford, CA 94305

Global analysis and differential geometry to JERRY L. KAZDAN, Department of Mathematics, University of Pennsylvania, E1, Philadelphia, PA 19104-6395

Probability and statistics to RICHARD DURRETT, Department of Mathematics, Cornell University, Ithaca, NY 14853-7901

Combinatorics and number theory to CARL POMERANCE, Department of Mathematics, University of Georgia, Athens, GA 30602

Logic, set theory, general topology and universal algebra to JAMES E. BAUMGARTNER, Department of Mathematics, Dartmouth College, Hanover, NH 03755

Algebraic number theory, analytic number theory and modular forms to AUDREY TERRAS, Department of Mathematics, University of California at San Diego, La Jolla, CA 92093

Complex analysis and nonlinear partial differential equations to SUN-YUNG A. CHANG, Department of Mathematics, University of California at Los Angeles, Los Angeles, CA 90024

All other communications to the editors should be addressed to the Managing Editor, DAVID J. SALTMAN, [redacted] at Austin, Austin, TX 78713.

General instructions to authors for

PREPARING REPRODUCTION COPY FOR MEMOIRS

> **For more detailed instructions send for AMS booklet, "A Guide for Authors of Memoirs." Write to Editorial Offices, American Mathematical Society, P.O. Box 6248, Providence, R.I. 02940.**

MEMOIRS are printed by photo-offset from camera copy fully prepared by the author. This means that the finished book will look exactly like the copy submitted. Thus the author will want to use a good quality typewriter with a new, medium-inked black ribbon, and submit clean copy on the appropriate model paper.

Model Paper, provided at no cost by the AMS, is paper marked with blue lines that confine the copy to the appropriate size.

Special Characters may be filled in carefully freehand, using dense black ink, or **INSTANT** ("rub-on") **LETTERING** may be used. These may be available at a local art supply store.

Diagrams may be drawn in black ink either directly on the model sheet, or on a separate sheet and pasted with rubber cement into spaces left for them in the text. Ballpoint pen is not acceptable.

Page Headings (Running Heads) should be centered, in CAPITAL LETTERS (preferably), at the top of the page — just above the blue line and touching it.

LEFT-hand, EVEN-numbered pages should be headed with the AUTHOR'S NAME;

RIGHT-hand, ODD-numbered pages should be headed with the TITLE of the paper (in shortened form if necessary).

Exceptions: PAGE 1 and any other page that carries a display title require NO RUNNING HEADS.

Page Numbers should be at the top of the page, on the same line with the running heads.

LEFT-hand, EVEN numbers — flush with left margin;

RIGHT-hand, ODD numbers — flush with right margin.

Exceptions: PAGE 1 and any other page that carries a display title should have page number, centered below the text, on blue line provided.

FRONT MATTER PAGES should be numbered with Roman numerals (lower case), positioned below text in same manner as described above.

MEMOIRS FORMAT

> **It is suggested that the material be arranged in pages as indicated below. Note: Starred items (*) are requirements of publication.**

Front Matter (first pages in book, preceding main body of text).

Page i — *Title, *Author's name.

Page iii — Table of contents.

Page iv — *Abstract (at least 1 sentence and at most 300 words).

Key words and phrases, if desired. (A list which covers the content of the paper adequately enough to be useful for an information retrieval system.)

**1991 Mathematics Subject Classification*. This classification represents the primary and secondary subjects of the paper, and the scheme can be found in Annual Subject Indexes of MATHEMATICAL REVIEWS beginnning in 1990.

Page 1 — Preface, introduction, or any other matter not belonging in body of text.

Footnotes: *Received by the editor date.
Support information — grants, credits, etc.

First Page Following Introduction – Chapter Title (dropped 1 inch from top line, and centered). Beginning of Text.

Last Page (at bottom) – Author's affiliation.